Naming the Local

HARVARD EAST ASIAN MONOGRAPHS 404

Naming the Local

Medicine, Language, and Identity in Korea since the Fifteenth Century

Soyoung Suh

Published by the Harvard University Asia Center
Distributed by Harvard University Press
Cambridge (Massachusetts) and London 2017

The Harvard University Asia Center publishes a monograph series and, in coordination with the Fairbank Center for Chinese Studies, the Korea Institute, the Reischauer Institute of Japanese Studies, and other facilities and institutes, administers research projects designed to further scholarly understanding of China, Japan, Vietnam, Korea, and other Asian countries. The Center also sponsors projects addressing multidisciplinary and regional issues in Asia.

Library of Congress Cataloging-in-Publication Data

Names: Suh, Soyoung, author.
Title: Naming the local : medicine, language, and identity in Korea since the
   fifteenth century / Soyoung Suh.
Other titles: Harvard East Asian monographs ; 404.
Description: Cambridge, Massachusetts : Published by the Harvard University
   Asia Center, 2017. | Series: Harvard East Asian monographs ; 404
Identifiers: LCCN 2016052194 | ISBN 9780674976962 (hardcover : alk. paper)
Subjects: LCSH: Traditional medicine—Korea—History. |
   Medicine—Korea—History. | Language and medicine—Korea—History. |
   Medicine and psychology—Korea—History.
Classification: LCC GN477 .S84 2017 | DDC 615.8/809519—dc23
LC record available at https://lccn.loc.gov/2016052194

Index by Alexander Trotter

♾ Printed on acid-free paper

Last figure below indicates year of this printing
26   25   24   23   22   21   20   19   18   17

To my parents, Seungnam Paik
and Daeyong Suh

Contents

Contents

Tables and Figures

Tables

Figures

Acknowledgments

I have received tremendous support from mentors, colleagues, friends, and family while working on this book project.

First and foremost, this project could not have even started without Charlotte Furth's warm encouragement. I gained many valuable insights by attending her graduate seminar and extended my intellectual endeavors under her supervision. She supported my rough Ph.D. dissertation proposal, which was just a concept at the time, and nurtured it by providing valuable comments and criticism. The perspectives I learned from Charlotte while completing the dissertation eventually developed into the backbone of this book. Charlotte's intellectual vigor has inspired me beyond description, and her kind and timely advice never fails to push me forward. I also wish to express great thanks to Shigehisa Kuriyama, whose scholarship I always long to emulate. His emphasis on brevity, clarity, and creativity has always challenged me to rethink what constitutes a good historical narrative. Through refined lectures and seminars, he taught me the effort a scholar should make to create a meaningful dialogue with audiences. Owing to his mentorship, I was able to deepen the themes examined in my dissertation.

As a postdoctoral fellow, I joined a research team led by Volker Scheid at the East Asian Sciences and Traditions in Medicine Research Center, University of Westminster, London in 2009. His enthusiasm for raising new questions about East Asian medical traditions has guided me to publish two articles, parts of which are incorporated in chapters 2 and 5. I deeply appreciate Volker's warm encouragement and detailed comments in completing those earlier works. My colleagues on the team, Eric Karchmer and Keiko Daidoji, brought in valuable analytical angles through which I now better understand the history of Korean medicine within the broader context of East Asia. Yi-Li Wu and Fa-ti Fan read an earlier version of the manuscript and provided valuable suggestions for revision. I have benefited greatly from Yi-Li's studies on the intersection between gender and Chinese medicine, and gained new perspectives from her latest comparative analysis of East Asian medicine. Fa-ti's numerous writings about the history of science and technology in East Asia challenged me to rethink the tension between the "universal and particular" and the "global and local" in East Asian contexts.

John Duncan supported this project when few scholars of Korean studies in the United States paid attention to the history of medicine. I am greatly indebted to his generous supervision at the University of California–Los Angeles. Benjamin Elman, who first supported my graduate study at UCLA, later encouraged me to attend the "Early Modern Asian Medical Classics and Medical Philology" workshops at Princeton University between 2011 and 2012. The insights I received from these workshops are partly incorporated in chapters 1 and 2. Owing to Vivienne Lo's advice, I was able to transform a dissertation chapter into a journal article. I benefited greatly from her experience and wisdom, which are reflected in the earlier versions of chapters 1 and 4. Marta Hanson gave me an opportunity to present my book project at the East Asian Studies Speaker Series held at Johns Hopkins University in 2014. The responses I received from the talk, and her own comments about chapter 1, shed new light on the topic. As the editor of a special issue of *Asian Medicine*, Marta also gave me valuable suggestions in completing the earlier version of chapter 2. Joining the "Medicine and Empire" workshop at the University of Chicago in 2012, I benefited from Susan L. Burns's research on medical advertisements in early twentieth-century Japan, which was incorporated in chapter 4.

I must acknowledge Dongwon Kim for his willingness to read my book proposal and other rough essays and, more broadly, for his passion for the history of East Asian science and technology. I have benefited greatly from his advice, not only for this book, but also for navigating many important and difficult issues surrounding the pursuit of a history of Korean medicine in Anglophone academia. Sun Joo Kim read a part of this manuscript, then reminded me that the devil resides in the details. She gave me many useful suggestions that helped to accelerate the production of this book.

I'd like to express my special thanks to Dartmouth College for its generous support of this project. In particular, the Junior Faculty Fellowship in 2014 and 2015 gave me necessary time away from teaching and other duties. The John Sloan Dickey Center for International Understanding sponsored the Manuscript Review Program in May 2014, which enabled me to invite a group of internal and external scholars to discuss this project. I am particularly thankful to those who willingly joined the program, then presented invaluable comments: Sienna R. Craig, Steven J. Ericson, Douglas E. Haynes, Sunglim Kim, Richard L. Kremer, Dennis C. Washburn, Christianne H. Wohlforth, and Projit B. Mukharji, who happened to visit Dartmouth at that time. Pamela K. Crossley also provided comments on an earlier version of chapter 5. My colleagues at the Department of History and the Asian and Middle Eastern Studies Program gave me heartfelt encouragement and pragmatic advice. Although a couple of grants supported earlier work on this book, I would like to highlight that without Dartmouth College's sponsorship, this book may never have come to fruition.

I thank two anonymous readers who reviewed the manuscript for the press. My editor, Robert Graham, at the Harvard University Asia Center showed warm support from the beginning and made substantial efforts to publish this book without delay. I appreciate Angela Piliouras and Ashley Moore at Westchester Publishing Services for their detailed copy editing. Natalie Reitano helped me greatly in many stages of revision. Eungyong Lee at Seoul National University has never said "no" whenever I asked her a question about Japanese sources. Ji Hee Jung willingly helped me transliterate Japanese references. Nien Lin L. Xie at the Dartmouth College library searched for archival sources for me.

Readers may notice that I rely heavily on Korean scholarship, whose works have not been widely circulated outside Korea. Trained in both Korean and Anglophone academia, I was able to observe the differences and strengths of each research tradition. I cannot name all the Korean scholars whose works have nurtured me, but I must acknowledge that their rigorous studies in Korean ensured the completion of this book. In particular, I am grateful to Yung Sik Kim, who first inspired me to explore the history of East Asian science, medicine, and technology.

I cannot thank enough the female graduate students and scholars I met at the Korean Research Institute of Women in the late 1990s. The rigorous seminars back then pushed me to continually question the relationships among language, identity, and marginality.

An earlier version of sections of chapter 1 was published as "Herbs of Our Own Kingdom: Layers of the 'Local' in the Materia Medica of Early Chosŏn Korea," in *Asian Medicine: Tradition and Modernity* 4, no. 2 (2009): 395–422. An earlier version of chapter 2 was published as "*Shanghanlun* in Korea," in *Asian Medicine: Tradition and Modernity* 8, no. 2 (2013): 423–57. An earlier version of sections of chapter 4 will be published as "Marketing Medicine to Koreans, 1910–1945" in *Imagining Chinese Medicine*, edited by Vivienne Lo (Leiden: Brill, forthcoming). An earlier version of chapter 5 was published as "Stories to Be Told: Korean Doctors Between Hwa-byung (Fire-Illness) and Depression, 1970–2011," in *Culture, Medicine, and Psychiatry* 37 (2013): 81–104. I thank the publishers who gave permission to use the earlier material as part of this book.

My greatest thanks go to my family. My mother, Seungnam Paik, and my father, Daeyong Suh, have always supported me, never doubting the value of my intellectual pursuits. Their faith in God in all circumstances of life has had an enduring impact on me. My three younger sisters, Eunyoung, Nayoung, and Donghee, always cheer me up with their unique sense of humor and passionate commitment to career and family. I cannot think of myself as a researcher and teacher without the inspiration I have gotten from my sisters. I am greatly indebted to my parents-in-law, Jungho Nam and Yunjin Lee. Their love and prayer has always encouraged me to explore the world in my own way. My husband, Sangwook Sunny Nam, has sacrificed a lot to walk with me, and provided exceptional support without which I could not have survived. My children, Sean and Yirang, have given me immense love, joy, and solace. Without them, my journey in writing this book could have been gray and dull. Many friends I have met in small, local Korean churches in the United States have helped me keep moving forward when words fail and a sense of futility prevails.

Note on Conventions

I use McCune-Reischauer romanization for terms and names in Korean. Chinese and Japanese are rendered in Pinyin and Hepburn, respectively. East Asian names are given last name first, without a comma. Authors of works in English are the exception.

I used digitalized court records of Chosŏn Korea. For the year, I relied on the online archive's translation. The lunar month and date in numbers follow the year according to the traditional order. For instance, "*Sŏngjong sillok*, 1478. 10. 29" implies the source was recorded in the ninth year of King Sŏngjong's reign, in the tenth lunar month, on the twenty-ninth day. The digital source also displays the corresponding reign period of the Chinese emperor.

All English translations are my own unless otherwise indicated. I translated the titles of Korean books and articles if no English translation was provided by the authors. If the authors provided their own translation and transliteration, I respected them. A certain degree of inconsistency was unavoidable. For instance, I rendered *fire illness* as *hwabyŏng*, yet it was often written as *hwa-byung, hwabyung,* or *hwatpyŏng* by other authors.

Last, in the List of Characters, some Chinese characters may appear twice, as I recognize both Korean and Chinese transliterations. For instance, I wrote both *men* and *mun* for 門 as the character appears in both Chinese and Korean texts.

INTRODUCTION

Hŏ Chun (1539–1615) lived an exceptional life as a physician in sixteenth-century Chosŏn Korea. Born the illegitimate son of an aristocratic official (*yangban*), Hŏ Chun was educated in the Confucian classics and became a man of erudition. He began his medical career at the Palace Clinic (Naeŭiwŏn) around 1570 thanks to a fellow high officer's recommendation. He gained the trust of King Sŏnjo (r. 1567–1608) by successfully treating the crown prince's smallpox in 1590, and the king further supported Hŏ Chun after he accompanied the king on the royal exodus from the palace to Ŭiju during the war with Japan (1592–98). After the war, Hŏ Chun took full responsibility for compiling a medical anthology following King Sŏnjo's wish to reform "the crude overuse and rough repetition" of medical texts circulated in China and Korea.[1] Although Hŏ Chun was briefly exiled after the king's death in 1608, the banishment gave him the necessary isolation to complete the project. Hŏ Chun submitted his compilation to the new king, Kwanghae (r. 1608–23), in 1610; the court published it in 1613. Given these scholarly and clinical achievements, the highest official rank, "senior first," was conferred on Hŏ Chun as a posthumous honor.[2] In Korean history, there are many physicians known for their excellent medical skills or for being a king's favorite. But it was rather uncommon to combine exceptional scholarly talent, medical skill, and royal support at the same time.

Hŏ Chun's exceptionalism is further exemplified by his choice of the word *tongŭi*, or "Eastern medicine," which represented a geographical distinction of a medical tradition. By highlighting Eastern medicine in the title of his masterpiece, *Tongŭi pogam* (Precious mirror of Eastern medicine, 1613), he justified Chosŏn Korea as a marginalized yet authentic location for medical pursuits. He argued, "Medicine is supposed to have its northern or southern distinction. Our kingdom is remotely situated in the East, and the way of pursuing medicine has never been stopped here. Thus, the medicine of our kingdom also deserves to be called Eastern medicine."[3] Just as those revered Chinese physicians Li Gao (1180–1251) and Zhu Zhenheng (1281–1358) were known as the representatives of "Northern medicine" and "Southern medicine," respectively, Hŏ Chun hoped his synthesis would gain a regionally differentiated recognition.

Hŏ Chun's appeal to Eastern medicine paralleled his efforts to document indigenous sources of medicine. Referring to earlier texts about local botanicals, Hŏ Chun registered many names of locally available materia medica and described their nature in comparison with herbs from China. Among the herbs that are used more than four times in the entirety of his prescriptions, 90 percent are composed of what was known as *hyangyak*, or "local botanicals."[4]

More significantly, Hŏ Chun wrote a few medical texts in vernacular Korean, aiming to reach wider audiences. With the dynasty's support, texts about pregnancy and childbirth, emergency aids, curing smallpox, and special prescriptions of medicines called *nabyak* were composed in Korean. Hŏ Chun's compilation represented the culmination of medical literature in Korean since the invention of the Korean alphabet (Han'gŭl) in 1446. The dynasty aimed to broaden medical governance through the textual production of medicine. In addition, Hŏ Chun's skillful synthesis, covering the full repertoire of Chinese originals available by the sixteenth century, demonstrated his competence in mastering the textual production of medicine from China. Deft selections, not a mere cutting and pasting of the originals, fleshed out the structure of his anthology of medicine. In terms of motivation, language, content, and interpretative framework, Hŏ Chun's work indeed seems to be a culmination of the indigenization of medicine in early seventeenth-century Korea.

No wonder that his *Precious Mirror of Eastern Medicine* has received more scholarly and popular attention than has any other premodern medical text in Korea. A variety of legends and folktales about Hŏ Chun provided an imaginative resource for a contemporary novel, which has struck a chord since its publication in 1990.[5] A television soap opera, which aired between November 1999 and June 2000, dramatized Hŏ

Chun's life and achievements, earning high ratings. Paralleling his popularity has been the conventional strategy of conjuring up Eastern medicine, which persistently stimulates cultural pride and commercial profit. In a time of international competition and professional contestation, Hŏ Chun and his identification with Eastern medicine serve well to delineate Korean people's view of traditional medicine as uniquely Korean and traditional.

As an icon of indigenization, however, Hŏ Chun simultaneously discloses layers of exogenous factors, through which the local attributes of medicine were established, modified, or even degraded. Primarily, it should be remembered that his *Precious Mirror of Eastern Medicine* was written in refined classical Chinese, which was the hegemonic language among the Korean literati until the turn of the twentieth century. The elegant writing style helped his work gain acceptance in eighteenth-century literary circles in East Asia; Japanese editions of the *Precious Mirror of Eastern Medicine* appeared in 1724 and then again in 1799, and seventeen Chinese editions were printed starting in 1763.[6] Conversely, his vernacular publications mostly remained less popular than the *Precious Mirror of Eastern Medicine* in his own time and even today. Although Hŏ Chun's recognition of the local attributes successfully complemented his encyclopedic composition, the elite registration of the native attributes did not necessarily pave the way for vernacular initiatives in medicine.

The clinical significance of the *Precious Mirror of Eastern Medicine* was not immediate. Those who sought the simplest, easiest, and most effective way of curing the sick viewed the *Precious Mirror of Eastern Medicine* as including unnecessary philosophical discussions and crude illustrations. King Chŏngjo (r. 1776–1800), for instance, ordered Hŏ Chun's compilation to be abridged, criticizing that "the theory [of the text] and rationale in prescription is too complicated, hence, the structure and examples are not concisely arranged."[7] It is not surprising that in late nineteenth-century Korea, one of the most popular and beloved medical texts was Hwang To-yŏn's (1807–84) *Pangyak happ'yŏn* (Compendium of prescriptions, 1885), which had a simple layout and presented concise formulas with the latest Chinese information about materia medica.[8]

Hŏ Chun's nuanced framework of Eastern medicine did not always gain support. It was not until the late nineteenth century that Eastern medicine reappeared in a medical text's title to reflect the geographical distinction with a sense of medical lineage. Although the contemporary nationalist framework depicts Hŏ Chun's Eastern medicine as part of the essential and uninterrupted heritage of Korea, many Korean physicians and scholars since the seventeenth century have pursued their own art of medicine, viewing Hŏ Chun's synthesis as a decent reference but not the only or ultimate measure of authority. Not all Korean physicians and scholars passionately embraced the identity of Eastern medicine, but they actively sought the latest medical resources from China for their own synthesis and abridgement. Chinese medical texts—including a variety of practical manuals and extracts of medical classics, secret manuals for successful pregnancy and childbirth, and specific prescriptions for epidemics and pediatric treatment—provided novel and diverse resources for readers in Chosŏn Korea.

More to the point, the material conditions of Chosŏn Korea since the seventeenth century made it difficult to rely solely on indigenous resources. For instance, some medicinal herbs indispensable in everyday prescriptions, such as licorice root, have hardly been naturalized in Korean soil. Although the state repeatedly prioritized local botanicals over herbs from China, rare and expensive Chinese medical materials, such as rhinoceros horn (*sŏgak*), ox bezoar (*uhwang*), musk (*sahyang*), and borneol (*yongnoe*) were continuously sought by those Korean consumers who could afford them via private and official trade. Even today, Koreans rely on China for most of their licorice root.[9]

Hŏ Chun's engagement with Eastern medicine, local botanicals, and Korean vernacular exemplifies how the boundary of the local was (re)established in relation to broader networks of scholarship and audience. Identifying indigenous attributes of medicine entailed connections and disconnections among knowledge, practice, and material entity, either physically or imaginatively. In navigating those multilayered junctures, more interestingly, questions about the individual or collective self have been raised. As my text will demonstrate, the supposedly particular nature of the body, soil, regimen, and medical lineage gained significance beyond the realm of medicine, thereby providing a source for conjuring the cultural and national identities of Korea.

By considering Hŏ Chun as a case, this book addresses the aspirations and limitations of registering the local in the existing configuration of medicine. Given that both advocating and resistant voices have constituted the categories of the local, we may question when and how the emphasis on the autochthonous in medicine arose or was banished. To what extent did linguistic, geographical, and cultural awareness of the local attributes in medicine affect clinical efficacy? If they did, can we interpret the expression of the irreducible difference in the locally situated body, disease, and healing as signaling a skepticism about the universal applicability of medicine? In a similar vein, has the Korean articulation of local distinctiveness in medicine entailed a quest for epistemological parity, either explicitly or implicitly?[10]

Interestingly, in Korean history Hŏ Chun connotes a pattern, not an exception. Under Japanese colonialism (1910–45), in the liberated South Korea of the 1970s, and in cosmopolitan Seoul in 2010, new voices revived his rendering of Eastern medicine. Although situated in different moments, these voices resonate by highlighting Korea's particular position in pursuing authentic medical practices and knowledge. They argue that without recognizing Korea's marginalized yet crucial situation in terms of language, culture, and geopolitics, no meaningful therapeutic innovations could ever be advanced.

The echoing voices surrounding Hŏ Chun's legacy have often questioned the credibility of the authoritative language with which to express native attributes. Trained in traditional medicine or biomedicine, physicians who have advocated Hŏ Chun's sensitivity also emulated his choice of the vernacular for their publications. Not only the content but also the style, tone, and linguistic medium of the local mattered. In later times, the alternate, hegemonic languages of medicine changed from classical Chinese, first to Japanese between 1910 and 1945, then to English as of the 1950s.

As has been shown in European translations of medical and scientific subjects from Latin into vernaculars, textual transformation in specific languages indeed represents deliberate choices. Vernacular composition, such as writing, editing, reading, copying, and circulating, reflects a series of conscious decisions to combine sociocultural identity with knowledge production. The indigenous attributes of medicine, via the Korean vernacular, have provided a source for physicians, scholars, and even patients to consider where we stand in relation to the wider world and who we want to be in the future in light of past experience and heritage.

Paralleling this willed particularism, furthermore, is an acknowledgment of unchangeable hierarchy. Again, translating Latin into vernacular languages demonstrates the enormous gap between a high-status language, usually Latin, and its counterparts, which usually have "lower prestige as a written language." Accordingly, the hierarchical relation assumes that the indigenous sphere of knowledge is in opposition to "something more learned, more conscious, more prestigious."[11] The long-ing for vernacularism entails a different awareness and manifestation of power relationships, eventually unfolding dissimilar modes of cultural and political beings.[12] As Sheldon Pollock puts it, "The term vernacular, for its part, refers to a very particular and unprivileged mode of social identity—the language of the *verna* or house-born slave of Republican Rome—and is thus hobbled by its own particularity."[13]

The Korean endeavor to register the indigenous elements of medicine rests on an irresolvable dualism: a willed decision to make the particular visible simultaneously faces "ever-increasing incommunication"[14] and marginality. Recognizing the indigenous in a particular linguistic form simultaneously defines the less privileged realm of knowledge in relation to the more prestigious and authoritative mode of learning. With the ad-vance of colonialism and nationalism in the early twentieth century, the dualism of the Koreanization in medicine became more explicit. No won-der the rejection of particularism was also alive. The longing for mastery of the privileged mode of language in medicine prevailed at the expense of the possibility of plural names and epistemological parity.

Compared to the European invention of the indigenous, the Korean aspiration toward the local exhibits more native initiatives compounded by self-conscious strategies and compromises. Knowledge about "indig-enous" or "primitive" people, botany, and medicine was established with the advancement of global colonialism and the professional ethnography originating in Europe. For instance, between the sixteenth and eighteenth centuries Portuguese explorers in Brazil sent systematic reports about South American fauna and flora and their medical efficacy to European intellectuals. The European conquest and colonization of sub-Saharan Africa in the nineteenth and early twentieth centuries established insti-tutional networks to authorize knowledge about native African agricul-ture, medicine, and racial physiology. The French report about Indian botanicals in the eighteenth century or the British engagement with Chi-nese gardens at the treaty ports in the nineteenth century enriched Euro-pean mastery of indigenous worlds. It is not surprising that increasing awareness of the local or the indigenous has been intrinsically intertwined with European expansion since the fifteenth century.[15]

These earlier studies on the invention of locality hinge mostly on the critical inquiry of the European construction of the Other. It was Euro-pean agents who named unfamiliar people, nature, and customs as "in-digenous" or "primitive," then documented their attributes for European audiences.[16] Accordingly, the dialogues, trades, and collaborations initiated by European actors were scrutinized. My analysis of the Korean manifestation of the local is more grounded in Korean agents and their initiatives in defining and circulating indigenous attributes for both domestic and international audiences. As Martin W. Lewis and Kären E. Wigen aptly point out, Korea, a country the size of Britain, has been largely ignored by Sinocentric metageography among Anglo-American audiences, yet the country has endured "with a language of its own and a long, intricate, and well-documented political, social, and cultural history."[17] Given Korea's enduring and sophisticated practice of documentation over millennia; the country's intimate and complicated relationships with neighboring regions, cultures, and nation-states; and the dynamic coexistence between traditional and biomedicine in the country, it is worthwhile to trace the Korean composition of indigenous medicine as a source of self-fashioning.

This book does not aim to reduce regional, linguistic, and cultural diversity into a category of national state. Although I draw on conventional terms such as "Korea," "Chinese," and the "universal," I object to any limited application of national, monolingual, and civilization-centered categories. The idea of the indigenous has been largely equated with nationality since the twentieth century, yet multiple origins of the indigenous and their connections with other localities over time define the major analytical framework of this book.

Each chapter traces the origins of a specific term. The birth, growth, transformation, fall, and even afterlife of each term represents an individual's or a group's aspiration to give a name to a particular thing or relation, thereby repositioning themselves within wider networks of people, material entities, and traditions of medicine.

In chapter 1, I examine the documentation of local botanicals, or *hyangyak*, during the Chosŏn dynasty (1392–1910). Fostering the idea of medicine best fitted for "our own kingdom," the dynasty sponsored a series of compilations labeled as "local botanicals." The classificatory arrangements used to map the local botanicals often overlapped and were not organized into a clear set of categories. Considering the traffic in herbs and medical texts across political and geographical boundaries and the extreme diversity of botanical names, shapes, and attributes, texts on local botanicals cannot be said to clearly show what belongs to a local "us" or a foreign "them." Instead, registering some names of botanicals in the Korean vernacular, textualizing the folk names of certain species, and publishing a series of books focusing on local botanicals reflected the sociocultural need of scholars during the Chosŏn dynasty to imprint motifs of the "local" on materia medica, thereby pursuing effective medical intervention, authentic knowledge, and a proper place of the self in the world.

Alongside the local botanicals, the textual label of Eastern medicine has best served the Korean need for cultural distinction against its Chinese counterpart since the seventeenth century. The imagined East as an appropriate location for Chosŏn's medical tradition legitimated elite physicians' positionality in producing medical knowledge and practice. Chapter 2 analyzes patterns of (re)arranging one of the most famous Chinese classics, Zhang Ji's (150–219 CE) *Shanghanlun* (Treatise on cold-damage disorders, ca. 196–220), by Korean physicians in conjuring an indigenous style of medicine. By taking the Korean accommodation of the canonical text as an example, this chapter elucidates the way Eastern medicine was fashioned, as well as its clinical and cultural ramifications. The evolving Korean compilation of the Chinese classic paralleled the changing meaning of Eastern medicine, which reflects the Korean urgency to establish their identity in medicine to compete with Chinese and Western medicine.

Chapter 3 draws on three groups of medics under Japanese colonialism as they became major agents in producing and consuming knowledge about distinctively "Korean" attributes of the Korean body: Korean doctors of traditional medicine, Korean doctors of biomedicine, and Japanese doctors as educators and administrators of biomedicine in early twentieth-century Korea. These groups modified the terms of traditional medicine and tailored images of biomedicine to articulate the specificities of the Korean body as a way of securing their social and cultural positions as medical professionals.

In particular, chapter 3 traces the effects of the Japanese imperial project in standardizing Korea as a locale for a global audience. After 1907, Japan emerged as one of the most advanced centers of knowledge production on the indigenous attributes of Korean bodies. Reports on fauna and flora, endemic diseases, hygienic conditions, and healing customs were collected and publicized in Japanese, establishing the explicit "Chosŏn (Korean)" category as a unit of analysis. Moreover, the Japan-led biomedical enterprise shaped the way Koreans reported their native attributes in vernacular Korean from the viewpoint of modern science. Discursive exchanges and institutional contestations among the three groups reveal the possibilities and limits of biomedicine in (re)defining the locality of Korean medicine in a colonial setting.

Whereas the first three chapters center mostly on elite publications of medicine, chapter 4 examines medical ads targeted at popular audiences. Analyzing the style and content of medical advertisements between the 1890s and 1930s, this chapter traces the evocation of the indigenous as a source of marketing. The complex mode of both enhancing and erasing the local attributes of health and illness illustrates the material and visual evolution of Korean locality in the early twentieth century.

After Korea was liberated from Japan in 1945 and the Republic of Korea was established in the south in 1948, doctors and policy makers debated the appropriate system of health care. Biomedicine was accredited, but traditional medicine piggybacked on the postcolonial nationalist agenda and gained some initiative. On September 25, 1951, the National Medical Services Law approved both traditional medicine and biomedicine in South Korea, from which the dual system survives to the present.

Given this milieu, chapter 5 analyzes the process of making *hwabyŏng*, or "fire illness," an internationally recognized term for a Korean emotion-related disorder since the 1970s. To index *hwabyŏng* as a valid condition within professional medical circles, Koreans drew on both the traditional idea of "constrained fire" and the *Diagnostic and Statistical Manual of Mental Disorders*' modern identification of "depressive disorders." By examining the research on *hwabyŏng* conducted by Korean psychiatrists and doctors of traditional medicine since the 1970s, this chapter demonstrates how conceptions of Koreanness in medicine have been inextricably tied to the correct positioning of Korea in postcolonial networks of medical specialists. The project of defining a uniquely Korean malady reflects the desire of medical professionals to make the indigenous meaningful, thereby creating a tool to increase its circulation and gain foreign recognition. Since the 2000s, studies of *hwabyŏng* have reflected the endeavor to extend the communicative sphere of medicine. The emphasis on Korean specificity has supported the vernacularization of medical dialogue as a source of healing. Denoting multiple meanings, *hwabyŏng* as a name of an illness has been (dis)assembled at the juncture of domestic and global trends in medicine.

Each chapter delves into a specific moment of medical innovation, yet five individual stories are woven together to reveal a *longue durée* in the Korean indigenization of medicine. The birth, decline, and afterlife of five terminologies, I argue, illuminate an irresolvable dualism at the heart of the Korean endeavor to define linguistic, geographical, and cultural specificities as indispensable elements of medicine. Partly due to Korea's geopolitical positionality and to the intrinsic tension of medicine's efforts to balance the local and the universal, Korean exertion to officially document the indigenous categories in a particular linguistic form had to constantly negotiate its own boundaries against the Chinese, Japanese, and North American authorities who had largely shaped the knowledge grid. Through a translinguistic and transtemporal approach, this book scrutinizes the extent of complexity with which Korean agents had to struggle to name their local attributes in medicine.

CHAPTER ONE

Local Botanicals, or *Hyangyak*

The Correct Name of Herb and Self

In 1478, Yi Kyŏng-dong (active 1470–1504) advised the king to fully promote *hyangyak*, or local botanicals, for health.[1] Yi Kyŏng-dong claimed that Koreans often went without effective medical treatment due to the rarity of Chinese herbs and the unreliable supply of foreign materia medica. To resolve this problem, he suggested that locally available herbs in every province be collected and examined for their medicinal efficacy. Furthermore, he argued that the state should compile texts about local botanicals and circulate them widely. This advice echoes the elites' growing attention to local botanicals as a medicinal entity and a subject of scholarly writing. Medical texts including *hyangyak* in their titles emerged in the thirteenth century and reached their zenith when the voluminous state project *Hyangyak chipsŏngbang* (Standard prescriptions of local botanicals, 1433) was completed with King Sejong's (r. 1418–50) support.

The advocates of local botanicals embraced a belief in the mutually beneficial relationship between health and the environment. The nearby products were presumed to be the best fit for managing health and illness. Accordingly, many court records about *hyangyak* during the fifteenth century highlighted the dissimilarity between Middle Kingdom China and Chosŏn Korea. For instance, Yi Kyŏng-dong argued, "The wind and soil of our kingdom is different from those of the Middle Kingdom, and the people's constitution is also dissimilar." In response, King Sŏngjong (r. 1469–94) further denied any medical efficacy of *tangyak*, or "Chinese herbs," for Koreans.[2] The rationale for using local botanicals sounded unequivocal.

In early 1479, however, Yi Kyŏng-dong expressed concern about books being published on local botanicals.[3] He cautiously hinted that the mere celebration of native herbs neither produced medical innovation nor adequately met the people's interest in materia medica. He reminded the king and his fellow officials that the ambitious court compilation of local botanicals pursued a generation before remained unpopular and was not frequently published; moreover, it had led to confusion:

> The *Standard Prescriptions of Local Botanicals* have already been compiled. However, [people] these days prefer *Hwaje* [Formulas of the pharmacy service] to the *Standard Prescriptions*; hence, this compilation is not in circulation. In addition, texts of materia medica edited by our people show only the name [of materia medica] but not the illustrations; thus, people hardly know [how to match names with actual herbs]. Therefore, please order this to be compiled again according to Chinese originals.[4]

Yi Kyŏng-dong succinctly pointed out that not all Chosŏn Koreans embraced the court's promotion of local botanicals. Instead, *Formulas of the Pharmacy Service* (1107), the state compilation of Song China (960–1279), was assumed to be authoritative and practical enough.[5] In addition, he ascribed the limited circulation of the *Standard Prescriptions of Local Botanicals* to its shortcoming of improperly matching names and herbs. Without illustrations, Chosŏn Koreans found it difficult to discern different species of the same name or same species with different names. To correct the problem, he suggested that the original Chinese texts of materia medica should be more thoroughly referenced.

Yi Kyŏng-dong's explicit support and subtle caveat hint at the complicated nature of textualizing local botanicals for Koreans. We may ask when and how the evocation of the local in medical texts emerged or earned disapproval. Given the desire to differentiate the "wind and soil of our kingdom" from those of the Chinese, to what extent did the existing Chinese texts of materia medica define the scope and pattern of Korean identification of local botanicals? Who were the target audiences, and to what extent did the state compilation gain popularity? In addition to authorship, sources, and circulation, Yi Kyŏng-dong's comment testifies that the core inquiry of *hyangyak* hinged on clarifying the mismatch between herbs and their names. Was the confusion in botanical nomenclature ultimately resolved by the state documentation of *hyangyak*? If not, how did Korean elites respond to the continued misuse and multiple renderings of herb names? More tellingly, we may also examine the linguistic convention of the *hyangyak* texts. Paralleling the newly invented Korean alphabet in 1446, did the enthusiasm for documenting local botanicals also pave the way for a vernacular initiative in medicine?

Considering these questions sheds new light on existing scholarship about *hyangyak*. It is well known that recorded traditional medicine in Korea began with sharing Chinese textual traditions. This beginning was not brought about by a single conceptual breakthrough but rather by a series of deliberate modifications, which modern historians have often used to demonstrate the distinctiveness of Korean medicine. Previous scholarship has underscored the rise of local botanicals and taken them up as a gauge for measuring the degree of independence and the extent of indigenization of Korean medicine. For instance, Miki Sakae (1903–92), who has provided foundational narratives on the history of Korean medicine from its antiquity to the twentieth century, argued that the "independence of medicine" and the "uniqueness of medicine on the peninsula" were at stake based on the idea that the native attributes of local botanicals illustrate the unique features of Korean medicine.[6] Kim Tu-jong (1896–1988), another well-known historian of Korean medicine, also highlighted the "independent" attributes of *hyangyak* texts. Following this trend, a recent study continues to draw on the native initiative or "Chosŏn" Koreanization as a major research framework. Local botanicals thus play a significant role in understanding what has been regarded as "Korean" about Korean medicine.[7]

The existing scholarship about *hyangyak* is unquestionably grounded in the current concept of the nation-state as a major analytical category. The search for a realm in which there is an "independence of medicine" maps the sovereign territory of Korea on medical knowledge and practice. Given the essentialized categories of Korea or China, a teleological development that conforms to the contemporary ethnic or cultural distinction is counted seriously. Unique achievements—not the complicated process of borrowing and modification—have shaped the major research questions. Furthermore, a simple link between a nation-state and its monolingual practice has prevented scholars from paying more attention to the multilingual composition of botanical nomenclature. Only limited attention has been paid to the Korean use of Chinese as transliteration, efforts to translate medical texts into vernacular Korean, or Korean elites' aspiration to communicate in classical Chinese at the expense of their native tongue. In the end, the anachronistic and monolingual approach to *hyangyak* fails to fully explore the possibilities and limitations of medical terminology in fashioning collective or individual identity.

My analysis of *hyangyak*, on the contrary, underscores the temporal composition of local botanicals, which emerged in the middle of linguistic, cultural, and political exchanges and compromises. In doing so, this chapter argues that evoking local botanicals does not merely reflect the native initiative in medicine. Instead, the rise and fall of *hyangyak* texts discloses Korean elites' ambivalent attitude to local botanicals: a manifested determination to value indigenous resources, accompanied by unresolvable confusion and ignorance in authorizing the native attributes of medicine. Such equivocation to a certain degree reflects the linguistic and cultural conditions of Chosŏn Korea, where the ordering of the domestic natural world was not always under the country's own control. The articulation of this equivocal attitude—not a celebration of the uniqueness of Korean local botanicals—sets the purpose of this chapter.

Hyang: Origins and Development

THE MEANING OF *HYANG*

It is unknown when the Chinese character *hyang* was first introduced to Korean medical literature.[8] Contemporary dictionaries primarily define *hyang* as the counterpart of *kyŏng*, the capital.[9] Accordingly, *hyang* refers to the countryside or to a local town when combined with other characters. In "local army" (*hyanggun*), "local academy" (*hyanggyo*), and "local clerk" (*hyangni*), *hyang* connotes a regional deviation from the center.

In Chinese classics, *hyang* also indicates the name of an administrative unit. Qin (221–206 BCE) and Han (206 BCE–220 CE) China defined ten *ri* as one *chŏng*, and ten *chŏng* constituted one *hyang*. Officials were assigned to each *hyang*, and the entire world under heaven was supposed to consist of 6,622 *hyang*. *Zhouli* (Rites of Zhou) views *hyang* as the largest unit of community; five households, or *ka*, are composed of *pi*, five *pi* become *ryŏ*, and in a similar vein, *chok, tang, chu,* and *hyang* were used to establish higher administrative divisions. All in all, 12,500 households belong to one *hyang*.[10] The earliest Korean use of *hyang* as a special administrative unit is found in the Silla kingdom (?–668 CE). *Hyang* indicated the forced-labor settlements of the lowborn, such as criminals, prisoners of war, and slaves. The less privileged *hyang* residents were banned from government positions and education, and their unfree status indicates the aristocratic attributes of the Three Kingdoms period. The *hyang* as an administrative unit for the less privileged survived in the Unified Silla (668–935) and Koryŏ (918–1392) kingdoms, but were mostly abolished by the early Chosŏn dynasty.[11] Either as a countryside, an administrative unit, or a sociogeographic differentiation, *hyang* in East Asian textual traditions presents a place-specific mapping of a governed world.

The meaning of *hyang* in medical literature can be inferred from another homonym, *hyang*, which signifies an act of facing something or countering. Chŏng Yag-yong (1762–1836), one of the most well known scholars in Korean history, describes this meaning of confronting *hyang* quite explicitly. According to Chŏng, *hyang* implies two vertical sides of the capital that face each other. Under a sage's governance, the capital was supposed to be divided into nine areas. The royal palace was centered in the middle, and the upper and lower divisions adjacent to the center were for the government office and the market. The remaining six areas were divided into three each on the left and right sides, respectively. This juxtaposition, or particular way of organizing the capital, was called *hyang*.[12]

Chŏng's comment warns us not to simply equate *hyang* with the countryside. Reflecting a place-specific idea, in elite documentation of medicine *hyang* was often designated as a bifurcation, or an act of setting up a binary. Although I translate *hyangyak* as "local botanicals," the "local" conjured up here seems meaningless without its counterpart, *tangyak*, "Chinese botanicals." *Hyangyak* hardly represents the regional variation on the Korean peninsula. Rather, it contends with "Chinese botanicals" as a countering concept.

The juxtaposition of *hyang* and *tang*, the domestic Korean and the foreign Chinese, was solidified particularly in the fifteenth century with the establishment of the Chosŏn dynasty. *Hyangyak* was expressed "in comparison with Chinese botanicals" or defined as being "dissimilar with the one produced in the Middle Kingdom."[13] For instance, Hwang Cha-hu (1363–1440) proposed publishing a medical text that was solely based on local botanicals, without relying on any Chinese botanicals.[14] *Hyang* in fifteenth-century medical discourse connotes a place-specific identification with respect to outward confrontation rather than inward differentiation.

THE LOCAL BEFORE *HYANGYAK*

Before the word *hyangyak* emerged in medical texts, the local attributes of botanicals in Korea were recorded with specific names of Korean kingdoms. This pattern of distinguishing botanicals was first found in the records of the Silla kingdom.[15] Today it is almost impossible to know the true extent to which Chinese texts circulated in Silla; nevertheless, medical texts, materia medica, and even famous medical practitioners moved between Silla and the Chinese territories.[16]

One of the most noticeable aspects of this traffic was the trade in herbs. *Samguk sagi* (History of the three kingdoms, ca. 1145) records Korean tributary offerings of ginseng and ox bezoar. Most ginseng from Silla was valued as high quality, as evidenced by comments in Liu Yu's (733–804) *Chajing* (Canon of tea, ca. 760). *Jiayou buzhu shennung bencao* (Supplementary commentary on the materia medica of the heavenly husbandman, 1061) also recognized the relatively high quality of ginseng from the Silla kingdom in the "Discourse on the Nature of Medicinal Drugs." Although the *Supplementary Commentary on the Materia Medica of the Heavenly Husbandman* was compiled long after the Unified Silla dynasty ended, ginseng from Korea was still referred to as "Silla ginseng."[17]

Silla ginseng, identified by its growing area, continued to be mentioned in Song materia medica, as evidenced in Su Song's (1020–1101) *Bencao tujing* (Illustrated classic of materia medica, ca. 1058–61) and *Bencao yanyi* (An exposition of materia medica), compiled by Kou Zongshi in the twelfth century.[18] Sources available now, such as *Jingshi zhenglei beiji bencao* (Materia medica, classified and verified from the classics and histories, ca. 1082) and *Haedong yŏksa* (History of Korea, ca. 1800), also recognized several kinds of botanicals as "Silla." "Silla beef tallow" and "Silla peppermint" were recorded in the *Illustrated Classic of Materia Medica* and in texts kept in the treasure repository of Shōsōin in Japan, respectively.[19] This distinction, made by identifying species and a particular source area, shows how botanicals from Korea gained a name and place within the Chinese classification system. Other than such examples, it is difficult to trace how Korean locality was mapped onto botanicals.[20]

In this flow of material goods and books, the latest Chinese texts and some rare materia medica were acquired by Koreans, while ginseng and ox bezoar were often offered to the Tang dynasty (618–907) court as a tribute. Chinese texts and some Japanese sources acknowledged Silla as a geographic classification mapping the diversity of botanicals. Still, Chinese sources mentioned in passing the regional variety of botanicals, without any exclusive emphasis on ethnic distinction. Considering the extreme diversity of plant names, shapes, and attributes, it seems more of an invention than a natural outcome to define what belongs to "us" and "them." In a similar vein, correcting the names of botanicals, employing the folk names of certain species in scholarly texts, and publishing a series of books focusing on local botanicals reflect a social and cultural need to articulate the local with respect to materia medica. To scrutinize this process, it is necessary to view the Koryŏ dynasty's publications about *hyangyak*. The medical writings during this dynasty provided the terms for and patterns of local botanicals through which writers of its successor dynasty, the Chosŏn, articulated and broadened their claims to the unique nature of local botanicals. The Koryŏ dynasty's interest in local botanicals and the following Chosŏn dynasty's articulation established two major texts, *Hyangyak kugŭppang* (Prescriptions of local botanicals for emergency use, ca. 1236) and *Standard Prescriptions of Local Botanicals*, as the foundation of Korean materia medica.

The Local in Koryŏ's Medical Texts

THE EMERGENCE OF LOCAL BOTANICALS

Koryŏ literature on local botanicals, although comparatively better documented than that of Silla, is still insufficient to fully illustrate the entire process of making local botanicals significant. Knowledge about Koryŏ's materia medica comes largely from one extant collection of prescriptions, *Prescriptions of Local Botanicals for Emergency Use*, and comments on medicine found in the *Koryŏsa* (History of Koryŏ, 1454) and *Koryŏsa chŏryo* (Essentials of Koryŏ history, 1452). In addition, a few titles recorded in later collections of prescriptions enable us to understand the increasing interest in local botanicals.

The existing sources indicate that a series of texts labeled as describing local botanicals emerged during the late Koryŏ dynasty. Table 1 summarizes bibliographical information on the publications that include the term *hyangyak*. Excluded from the table are a couple of texts without the term *hyangyak* in their titles that are also known to have contributed prescriptions to the later compilations of local botanicals.[21]

What explains the increased amount of *hyangyak* literature during the late Koryŏ dynasty? Although there is no single answer, scholars have

Table 1. Major Texts Labeled as Describing Local Botanicals
during the Koryŏ Dynasty

Title	Date	Author	Availability
Prescriptions of Local Botanicals for Emergency Use (*Hyangyak kugŭppang*)	ca. 1236	Unknown	One edition preserved in Japan
San Hezi's Prescriptions Using Local Botanicals (*San Hezi hyangyakpang*)	?	Unknown	Lost; title mentioned in the preface of *Standard Prescriptions Using Local Botanicals for Universal Salvation* (*Hyangyak chesaeng chipsŏngbang*, 1399)
Old Prescriptions Based on Local Botanicals (*Hyangyak kobang*)	?	Unknown	Lost; title and four formulae found in *Standard Prescriptions of Local Botanicals* (1433)
Tested Prescriptions Using Local Botanicals to Benefit the People (*Hyangyak hyemin kyŏnghŏmbang*)	?	Unknown	Lost; title and twenty-six formulae found in *Standard Prescriptions of Local Botanicals* (1433)
Easy Prescriptions Using Local Botanicals (*Hyangyak kanibang*)	?	Unknown	Lost; title and fifty formulae found in *Standard Prescriptions of Local Botanicals* (1433)

SOURCES: Drawn from the descriptions in Miki, *Chōsen igakushi oyobi shippeishi*, 67–69; Kim Sin-gŭn, *Han'guk ŭiyaksa*, 52–56; and Kim Tu-jong, *Han'guk ŭihaksa*, 139–40, 153.

conjectured that the Koryŏ elites' confidence in consuming medical texts from China, official and private exchanges of herbs, and the idea of helping people by expanding their medical knowledge motivated the documentation of indigenous herbs.[22]

Foreign materia medica were introduced to the Koryŏ court by merchants and envoys from China, Japan, and Arabia. These visitors offered a variety of rare and precious goods. In return, Koreans offered Song China ginseng, pine nuts, and fragrant oil. The list of imperial gifts presented to King Munjong (r. 1046–83) by the Song emperor includes the names and origins of more than one hundred Chinese materia medica. Described as "clove of Guangzhou," "crow dipper of Jizhou," and "bupleurum of Yinzhou," these materials signified imperial benevolence to a neighboring Korean king, who was suffering from paralysis caused by a stroke.[23] As a response, the Korean king reciprocated by sending ginseng.

Medical experts were consulted across political boundaries. For instance, King Munjong asked for advice on matters of his personal health. According to five sets of records in *History of Koryŏ*, medical experts from Song China were mostly concerned with treating Munjong's numbness. Song physicians accompanied envoys, transmitting medical texts and herbs. Xin Xiu (?–1101), who was naturalized as a Koryŏ subject during Munjong's reign, was known for his medical knowledge even though he was not a medical official. Another Chinese scholar, Xu Jing (1091–1153), accompanied the envoys dispatched to Koryŏ during the reign of Song emperor Huizong (r. 1100–1126) and wrote of his experience in the Koryŏ capital in a book titled *Xuanhe fengshi gaoli tujing* (Illustrated classic of Koryŏ written by the envoy of the Xuan Huo period, 1123). Composed of twenty-nine sections (*men*) and three hundred entries (*tiao*), this account includes a few comments on the customs and medical conventions of Koryŏ, such as the tea ceremony, the preservation of ginseng, and domestic trade in medicine.[24]

An increase in the diffusion and circulation of texts paralleled the flow of medical experts and the exchange of materia medica. Extant records testify to the Koryŏ court's publication and transmission of Chinese medical texts. Song medical texts, such as *Taiping shenghui fang* (Taiping era formulary of sagely grace, 992), seem to have been imported by official envoys in 1016 and 1022. During King Munjong's reign, this kind of intermittent textual importation was replaced by orderly, planned state publications. Sources reveal that medical texts were published in Korea in 1056, 1058, and 1059, including classics such as *Huangdi bashiyi nanjing* (The Yellow Emperor's canon of eighty-one problems, ca. second century) and the *Treatise on Cold-Damage Disorders*.[25] Those classics were not first introduced to Koryŏ during these years; their publication in this time reflects the increasing demand for these texts. Certainly, Chinese medical literature was regarded as more than just an emblem of diplomatic recognition. The official Korean publication of medical texts reflects the level of cultural pride and in-depth knowledge of the major Chinese medical texts during the Koryŏ dynasty. The Song emperors were aware of Koryŏ's competence: Emperor Zhezong (r. 1085–1100), for instance, ordered his envoy to Koryŏ to look for fine woodblock copies of eleven medical texts.[26] Although only one, *Huangdi zhenjing* (The Yellow Emperor's classic of acupuncture), was actually sent to the Song rulers from the Koryŏ court, this list identifies possible texts the Koryŏ dynasty might have added to the court's collection.[27]

To sum up, the emergence of *hyangyak* texts reflects Koryŏ's gradual digestion of medical literature from Song China and frequent exchanges of herbs via trade and diplomatic missions. Confident in their medical learning, Koryŏ Koreans aimed to correct the Chinese sages' "multitudinous" teachings and extract essential prescriptions to effectively treat both emergency and chronic cases.[28] Given the revisionist approach to Chinese texts, it is a small wonder that the conventional wisdom of folk medicine and locally available herbs was viewed in a new light. The extant preface of *Piye paegyobang* (One hundred essential prescriptions reserved, ca. 1270–90), which emulated the *Prescriptions of Local Botanicals for Emergency Use*, argues that medical solutions are derived from mundane entities. The preface highlights that to a modest and prudent mind, everything in the world may have medical virtue.

> Everything between heaven and earth can be used as medicine. Mountain, hill, farm field, castle, road, garden, stone steps, room, gate, wall, clothes, bedclothes, broken bowl, rotten things, everything from soup to nuts. Therefore, [if you contract] any disease, [you should remember] medicine is already in preparation. If you are careful, like crossing deep water or stepping into shallow ice, and do not ignore serious matters, then how can even dust not reveal its medicinal virtue?[29]

Medical solutions did not arise from special preparations. According to the preface, the surrounding environment and conventional things of life already have curative properties. As such, it is worthwhile documenting indigenous herbs and using them for therapeutic solutions. Koryŏ's texts about local botanicals are mostly lost, only being cited by their titles by later publications. The authors' own voices, the scope of circulation, and the impact on clinical encounters remain largely unknown. The extant preface reveals at least the influential framework through which local botanicals gained validity. The Chosŏn's *hyangyak* texts in the following centuries repeat this idea that medical solutions already exist in mundane things. Interestingly, however, the earliest extant Koryŏ preface did not contrast *hyang*, the local, with *tang*, the Chinese, as explicitly as later Chosŏn authors did. Before we examine the evolution of the *hyangyak* texts in two dynasties, the content and style of the documentation of the local in Koryŏ should be explored.

LAYERS OF THE "LOCAL"

Prescriptions of Local Botanicals for Emergency Use is one of the oldest known texts on local botanicals in Korea. Furthermore, this text is viewed as having provided the thematic framework and content for the authors of the later *Standard Prescriptions of Local Botanicals*, which is one of the most significant medical texts about local botanicals in the early Chosŏn dynasty. Not surprisingly, our knowledge about the organization and content of *hyangyak* literature comes from the (dis)continuities between these texts.

The structure of *Prescriptions of Local Botanicals for Emergency Use* employs the organizational system of Tao Hongjing's (456–536) *Bencao jing jizhu* (Collected commentaries on classical materia medica of the heavenly husbandman, ca. 530–57), which sets up three layers of drug quality, to arrange the materia medica. Likewise, *Prescriptions of Local Botanicals for Emergency Use* is composed of three volumes (*kwŏn*), and each volume is classified according to the upper, middle, and lower divisions.[30] Alongside these sections, a list of local botanicals used in prescriptions, "Pangjung hyangyak mok ch'o bu" (List of local botanicals), presents 180 species.[31] Here the local name, attributes, toxicity, and method of collecting each herb are described.

In what ways does this list define the local attributes of botanicals? Availability primarily shapes the boundary of *hyangyak*. Accordingly, not all of the 180 species in the list are plants uniquely native to Korea. For instance, ginseng, iris (*ch'angp'o*), *Liriope platyphylla* (*maengmundong*), *Euphorbia pekinensis* (*taegŭk*), and ox bezoar were included in the plants of the *hyangyak* list, but they were already well known in the texts of Chinese materia medica. Herbs are classified as "local" because they were grown in Korean soil, but they were not exclusive to Korea.

More to the point, the species categorized as "local botanicals" were identified with a folk name (*sogun*), written in Chinese characters yet pronounced differently from the Chinese. Known as the *idu* system, this method employs Chinese characters to signify Korean pronunciation arranged in the Korean word order.[32] 松衣羅 is thus the *idu* native name for a species like the iris (*ch'angp'o*), 冬沙伊 is *idu* for *Liriope platyphylla*, 數板麻 is *idu* for *Astragalus membranaceus* (*hwanggi*), and 狄小豆 is *idu* for *Cassia tora* (*kyŏlmyŏngja*).[33] Not all species were designated by *idu* names. Some already well-established herb names, such as ginseng, ox bezoar, *Rehmannia glutinosa* (*chihwang*), and chrysanthemum (*kukhwa*), were based on their Chinese names. In addition, all the information about the species was recorded in classical Chinese.

The organization of *Prescriptions of Local Botanicals for Emergency Use* reveals multiple additions and editorial revisions. In addition to the herbs recorded in the "List of Local Botanicals," more than forty plants that are not registered in the list, such as licorice and ginger, are found in the prescriptions of *Prescriptions of Local Botanicals for Emergency Use* to which the list belongs.[34] Among them are included plants that were commonly found in Korean soil, such as ginger and mustard.

To summarize, the list incorporated in *Prescriptions of Local Botanicals for Emergency Use* includes some local botanicals that were not native to Korean soil. Korean pronunciation of herb names was recognized, but only as complementary. The majority of well-established herb names and information on them were recorded in classical Chinese. Some botanicals that were excluded from the list but included in the prescriptions were found in the local habitat. What, then, was the primary standard used to classify these species as "local" herbs? The boundaries of this category remained ambiguous.

As its title suggests, *Prescriptions of Local Botanicals for Emergency Use* was compiled for the convenience of people who had difficulty reaching good doctors and obtaining proper treatment suggestions. Using plants that are locally available, *Prescriptions of Local Botanicals for Emergency Use* aimed to provide a practical tool for educated elites managing health and disease. Given this milieu, the intrinsic inconsistency in the way botanicals were considered local might not have been the main concern of the text's editors. Even so, the way local botanicals were committed to scholarly texts should be scrutinized, because *Prescriptions of Local Botanicals for Emergency Use* has been viewed as a significant step toward the "independent development of our own medicine" and the "self-management of medical policy."[35] Furthermore, the continuous tradition of local botanicals that began with this book resulted in the establishment of Korea's own study of materia medica during the Chosŏn dynasty, thereby contributing to the growing sensitivity toward indigenous medicine.[36]

As discussed earlier, the use of the term "local" documented in *Prescriptions of Local Botanicals for Emergency Use* posits an indigenous nomenclature, yet the native names themselves were not standardized. Some folk names, presented in *idu* in both the *Prescriptions of Local Botanicals for Emergency Use* and the "List of Local Botanicals," did not match. For instance, angelica (*tanggwi*) was rendered as an *idu* word, 黨歸菜根, in the *Prescriptions of Local Botanicals for Emergency Use*, whereas angelica was written as 且貴草 in *idu* in the "List of Local Botanicals." *Angelica polymorpha* (*kunggung*) was presented as 芎芎草 in *idu* in the prescriptions and listed as 蛇休草 in *idu* in the list. If that is not confusing enough, whereas the "folk name" or "commonly said" (*sogun*) name was displayed in the list, the "local name" (*hyangmyŏng*) was used in the prescriptions.[37]

The incongruity revealed in the naming of local botanicals illustrates the complexities involved in defining local botanicals. Most well-known materia medica, such as ginseng and ox bezoar, did not need a local name. When 180 species were categorized as "local botanicals," the implication was that the botanicals were locally available, not necessarily locally grown. Some included in the prescriptions were apparently supplied by foreign trade.[38] Even herbs that were seemingly popular and generally adapted to Korean soil had a variety of names based on the growing area and production.

Finally, it should be noted that when these local botanicals were committed to scholarly texts, the scholarly apparatus still referenced major Chinese texts. The "List of Local Botanicals" in *Prescriptions of Local Botanicals for Emergency Use* relied on *Collected Commentaries on Classical Materia Medica of the Heavenly Husbandman* for information on herbs' morphologies and seasons and on *Illustrated Classic of Materia Medica* for their general attributes, toxicities, and medicinal effects. Thus, it is plausible to surmise that *Prescriptions of Local Botanicals for Emergency Use* was compiled by referring to the latest Chinese text on materia medica while adding the folk names of some of the available local botanicals.[39]

Certainly, Koryŏ's compilation of local botanicals reflects the growing interest in using native resources for cures. As *Prescriptions of Local Botanicals for Emergency Use* exemplifies, documentation of the names of locally available herbs is distinguished from other replicas of Chinese texts published in the earlier period. However, as was shown in the previous section, the "local" was recognized and presented as textual knowledge without any clear-cut classificatory scheme about local botanicals. Through the foregoing discussion, we can see that the authors of these materia medica were using multiple definitions of "local" when referring to the identity of botanicals: first, those available to Koreans but possibly imported; second, botanicals grown locally but also found in other places; and third, botanicals with local or folk names but still referred to in classical Chinese as a major linguistic medium. These three perceptions of "local" identity often overlapped in practice and were hardly organized into a clear set of catalogs.

The Koryŏ's textual composition of the local evolved. A combination of *hyang*, the Chosŏn Korean, and *tang*, the foreign Chinese, came to be imposed on varieties of herb names. This process of creating "our own" botanicals necessitates consideration of additional ideological and cultural catalysts promoted by the new political regime of the Chosŏn dynasty.

Local Botanicals in the Early Chosŏn Dynasty

MATERIA MEDICA IN MOTION

During the early Chosŏn period, medical experts, medical literature, and materia medica were all in vigorous circulation. Many records in *Chosŏn wangjo sillok* (Veritable records of the Chosŏn dynasty) indicate that tributary envoys to Ming China (1368–1644) acted as a major route for the herb trade.[40] As in Koryŏ, dozens of medicines, such as ginseng, pine nuts, and deer horn, were often offered by the Chosŏn court to Ming China, whereas ox bezoar, borneol, cinnabar, pepper, and frankincense were listed as materials sent from the Ming court to Chosŏn.[41] Although the exact scale of exchange is not entirely known, tributary offerings ranged in the hundreds of *kŭn* per item,[42] and private trade, undertaken during envoy trips, flourished. Memorandums to the throne often urged state intervention to correct nefarious conventions in the herb trade and disclosed various problems that arose as a result of the ambiguous borders between official and private trade. However, trade associated with envoy trips could not be rigidly controlled by state regulations.[43]

Ginseng best represents the diplomatic and commercial interests associated with the exchange of herbs. Court records of the late Koryŏ and the early Chosŏn dynasties indicate the high demand of ginseng for the tributary offering to Song, Yuan, and Ming China; in 1071, "Kim Che was sent to Song with 1000 *kŭn* of wild ginseng";[44] in 1279, "Yi Paek-ch'o was sent to Yuan with ginseng."[45] Koreans began to cultivate ginseng in the eighteenth century; before then, all ginseng was referred to as wild ginseng collected from forests. Needless to say, the increased court demands for ginseng incurred people's resentment. By the end of the thirteenth century, the *History of Koryŏ* stated that "the king Ch'ungnyŏl's queen, the princess of Yuan Empire, had good profits by sending ginseng and pine nut to the Jiangnan China. Thus, she later sent eunuchs to every corner of the country, even where no ginseng and pine nut were produced, then collected them in large quantities. The people suffered enormously."[46]

The demand for ginseng hardly decreased in the Chosŏn dynasty. The collection of ginseng as a general tribute to the state was incumbent on every province. In the late fifteenth century, the amount collected nationwide reached more than 1,800 kg (3,000 *kŭn*). From the northernmost Hamgyŏng to the southernmost Kyŏngsang provinces, no region was exempt from this authoritative collection by the state.[47]

The state prohibited ginseng trade across the borders. The envoys to Ming China, however, were allowed to carry certain amounts of ginseng and trade them in Beijing to meet the cost of diplomatic travel. Gradually, in the fifteenth century, officials and translators discovered the increasing profits of ginseng, and, accordingly, the private trade, which was originally authorized for the purpose of diplomatic exchange, became the source of expanded commercial activities, smuggling, and illegal transactions. In the early seventeenth century, Korean merchants specializing in ginseng emerged, and ginseng constituted 20 percent of all Japanese imports from Chosŏn Korea in 1684. Ginseng commanded the major flow of Japanese silver to the peninsula throughout the seventeenth century.[48] Given this commercial potential, ginseng had never been completely controlled by the Chosŏn dynasty's legal code.

Linda Barnes argues that ginseng was one of the two most important and profitable drugs in Sino-Western trade beginning in the eighteenth century.[49] Even before then, ginseng crisscrossed regional and political boundaries in East Asia, serving as a source of revenue and as a diplomatic emblem. Did the increasing significance of Korean ginseng have any impact on the rise of *hyangyak* in medical texts? No explicit correlation is found. Still, court records about controlling ginseng and boosting local botanicals coincided. The bulk of these records are found during the reign of King Sejong, who supported the grandiose compilation of the *Standard Prescriptions of Local Botanicals* and aimed to extract government revenues from ginseng.

Paralleling the herb circulation, medical experts also crossed dynastic boundaries from the late fourteenth through the sixteenth centuries. Chosŏn envoys often asked the Ming emperor to send medical experts, and many Korean medical officials who accompanied official envoys tried to acquire more texts, materia medica, and knowledge from Ming physicians. Some texts actually enable us to reconstruct this dialogue between Ming scholars and Chosŏn medical officials. *Chungjo chilmunbang* (Sino-Korean questions about prescriptions, ca. 1506–43), *Chungjo chŏnsŭppang* (Sino-Korean prescription transmission and practice, ca. 1506–43), and *Tap Chosŏn ŭimun* (Answers to the questions of a Korean physician, 1624) are known to present-day scholars, although not all of the texts have survived. The fact that there is a Japanese edition of *Answers to the Questions of a Korean Physician* tells us the range of this text's circulation during the eighteenth century.[50]

Among the aforementioned titles, *Answers to the Questions of a Korean Physician* transcribed the responses of Wang Yinglin (active 1610s–20s), a Chinese official in the Ming dynasty, to the questions of Yun Chi-mi (active 1591–1612), a Korean official physician. Composed of twenty-four articles and written in classical Chinese, the text reveals issues ranging from understanding unresolved clinical symptoms to elaborating on details of the body's circulation channels and arriving at specific therapies for diseases.[51] Transcribed in the early seventeenth century, the text exemplifies the way the encounters of medical specialists across dynastic borders became a method of furthering medical knowledge. The experts' encounters did not merely remain in the realm of diplomatic formality but generated paths through which the latest interpretations of diseases, novel prescriptions, and empirical therapies were transmitted via interpersonal communication.[52]

Paralleling this flow of material trade and interpersonal encounters was an upturn in the appeal to "local botanicals" or "locality" as authoritative in textual forms of knowledge. This upturn was consonant with a more enhanced claim for a medicine that was supposed to originate from native soil. The Chosŏn dynasty, as the sponsor of the grandiose state project, evinced its special interest in cataloging the local, thereby conjuring up the Chosŏn-centered art of medicine. However, the clinical impacts of the top-down approach on medical innovation remain unclear.

THE STATE'S IDENTIFICATION
OF LOCAL BOTANICALS

The Chosŏn dynasty relied on medical texts mostly transmitted from China, which date back hundreds or even thousands of years. In applying the prescriptions from the ancient texts to the contemporary clinical encounters in Chosŏn, it was crucial to identify the names, shapes, and attributes of botanicals in the texts and compare them with locally available Korean herbs. Confusion in identifying botanicals was a significant issue in medical treatment. Some botanicals of the same species had different names, and different species were sometimes called by the same name. These circumstances complicated the official documentation of local herbs and their medical effects.

During King Sejong's reign, medical officials were dispatched to China to resolve confusions in herb names. In 1423, for instance, envoys to Ming China examined sixty-two Korean species in collaboration with Chinese scholars. In 1430 and 1431, Sejong's officials attempted to discriminate between the qualities of domestic botanicals, hoping to substitute them for the Chinese botanicals that were named in prescriptions but not available in Korea.[53] The investigation in 1423 concluded that fourteen domestic botanicals, including *Machilus thunbergii* (*hubak*), *Salvia miltiorrhiza* (*tansam*), and *Bupleurum falcatum* (*siho*), turned out to be different from Chinese products and were forbidden for use in medical preparation.[54] Because the referential standard resided in Chinese texts, only the domestic herbs that were identical to the Chinese originals were allowed to be used in Korea.

The exact process of identification is not known, yet the aforementioned examples demonstrate the complicated nature of defining usable local herbs for Koreans. Today's taxonomy does not differentiate Chinese *Salvia miltiorrhiza* from the Korean one. Known as red ginseng in general, the root of *S. miltiorrhiza* was widely used in China and Korea and had been deployed with many names. In addition, the *Bupleurum falcatum* had been recognized with multiple names even in Chinese materia medica. Northern and southern distinctions were marked regarding the name of *B. falcatum*. Known for boosting liver function and calming emotional discomfort, *B. falcatum* was indispensable to many prescriptions in major medical texts, such as the *Treatise on Cold-Damage Disorders*. Given the diversity of botanicals and plurality of herb names, the boundary between *hyangyak* and *tangyak* was not always obvious.

In conjunction with their identification, the dynasty made efforts to localize foreign botanicals, particularly those in high demand. Some remarks in the *Veritable Records of the Chosŏn Dynasty* reveal an attempt to grow licorice. For instance, King Sŏngjong encouraged the localization of licorice, which became more than just an ephemeral project. *Yejŏn chamnyŏng* (The miscellaneous ordinance of the code of rites, ca. 1492) informs us that during the early Chosŏn dynasty, the local cultivation of licorice was directed by officials dispatched from the Palace Clinic. The officials were expected to check the cultivation conditions in every province at least once a year. Not only licorice but also pepper (*hoch'o*), *Ephedra sinica* (or Chinese ephedra, *mahwang*), and *Alpinia galanga* (*yanggang*) were plants that Koreans tried to naturalize in their home soil.[55]

Aiming at enhanced governance, the state also made systematic efforts to fully document domestic botanicals, which was demonstrated by the official publication of *P'alto chiriji* (Geographical gazette of the eight provinces, 1432).[56] This detailed information was deemed necessary to govern local provinces. The gazette was commissioned to record the administrative history of each local government jurisdiction and its topographical features, control checkpoints, fortifications, land area, population, native products, roads and post stations, garrisons, troop levies, beacon communication sites, mausoleums and tombs, surnames found in the area, and historical personages.

Under the category of native products in the *Geographical Gazette of the Eight Provinces*, a section called *saengsan yakchae* (materia medica in production) listed all botanicals growing in each province. Cumulatively, the eight provinces were said to produce 1,080 separate plants, which were categorized into approximately 317 species. In the category of *chongyang yakchae* (materia medica in cultivation), dozens of botanicals under cultivation were delineated according to the administrative units. Finally, the category of *chinsang yakchae* (materia medica for offering to the king) was added to specify the items and quantities of materia medica that should be sent to the capital. Beginning during Sejong's reign, an offering to the king was distinguished from a general tribute in the legal code, thereby enhancing the court's initiative in managing materia medica. Under this category, herbs should be separately managed and reserved for one of three major medical offices in the capital: the Palace Clinic, the Directorate of Medicine (Chŏnŭigam), or the Office of Benefiting the People (Hyeminsŏ). If there were enough stored herbs to meet annual supply needs, the rest was distributed for general use.[57]

The Chosŏn dynasty's initiative in managing materia medica became noticeable in the fifteenth century. The state aimed to further control the supply of domestic botanicals: the medical efficacy of available plants was examined to replace Chinese originals in prescriptions; the naturalization of some species was supervised by the court; and the production and cultivation of plants in each province was carefully recorded.[58] In tandem with the state's overall interest in managing domestic herbs, the most grandiose medical compilation of *hyangyak* was carried out.

The state's involvement leads us to ask whether local botanicals, as material entities or forms of knowledge, had any effect on the conditions of ordinary people. Would the state's project of encouraging the production of local botanicals result in real medical effects, such as reducing infant mortality or promoting population growth? Scholars have reasoned differently. Yi T'ae-jin, for instance, ascribes the gradual increase in the population from the late Koryŏ to early Chosŏn dynasties to an effective state policy carried out in medicine and agriculture. He believes that local botanicals played a more significant role than agricultural innovation in reducing infant mortality, thus ultimately increasing the growth of population. But Sin Tong-wŏn discounts the role of medicine in population growth, referring to Thomas McKeown's argument on the relationship between population growth and Western medicine.[59] Taking into account factors such as the overall progress in nutrition and the development of an improved immune system, McKeown argues that Western medicine had a limited role in influencing population growth. In a similar vein, Sin Tong-wŏn notes that Yi T'ae-jin has not fully demonstrated the actual connection between local botanicals and their clinical impact on certain diseases that were supposedly fatal to infants.[60]

As it appears in scholarly discussion and existing records, the state documentation of local botanicals better demonstrates its intention, not its lasting impact. *Hyangyak* in official records reflects the court's will to expand medical benevolence as a part of Confucian governance, yet the top-down efforts to promote local botanicals did not always accompany medical innovations. Even so, the state's dedication to *hyangyak* deserves our attention. The Chosŏn's compilation displays the elites' speculation about desirable medicine and offers a rationale for the study of local botanicals that was repeated in later medical publications.

LOCAL BOTANICALS AS
A MANIFESTATION OF IDENTITY

Standard Prescriptions of Local Botanicals represents a high point in the history of Korean medicine. The official compilation of this title with another well-known medical text, *Ŭibang yuch'wi* (Classified compilation of medical prescriptions, 1477), makes a vivid display of the intellectual authority and cultural confidence of the Chosŏn dynasty. Both books were compiled during King Sejong's reign, at the same time as the implementation of one of the most sophisticated cultural projects in the entire Chosŏn dynasty. The extensive collection and modification of information was not limited to the field of medicine but was evident across a range of other topics, including astronomy, agriculture, and technology.[61]

Standard Prescriptions of Local Botanicals is composed of eighty-five volumes (*kwŏn*) in thirty books (*ch'aek*). The bibliography of *Standard Prescriptions of Local Botanicals* lists more than 160 medical texts, mostly transmitted from China. Made up of 57 major (nosological) categories (*taegangmun*) and 959 subsections (*mok*), *Standard Prescriptions of Local Botanicals* stood as the emblem of learned medicine. Although based on the major texts of Chinese medicine, *Standard Prescriptions of Local Botanicals* nevertheless stimulated knowledge of local botanicals by incorporating most of the Koryŏ's own texts on local botanicals, providing a list of previously discovered local botanicals, and suggesting some prescriptions with local herbs. The 959 subsections in the new compilation were a substantial increase from the 338 listed in the earlier court compilation, *Hyangyak chesaeng chipsŏngbang* (Standard prescriptions using local botanicals for universal salvation), in 1399. In addition, the previously known 2,803 formulas from the *Standard Prescriptions Using Local Botanicals for Universal Salvation* were expanded to 10,706, with 1,416 newly added items about acupuncture, in *Standard Prescriptions of Local Botanicals.*[62]

There is no doubt that the Chosŏn dynasty intended to reflect its cultural confidence and sovereign pride in the knowledge of materia medica. *Standard Prescriptions of Local Botanicals* was supposed to elaborate on why "we" should value "our" own medical resources. Kwŏn Kŭn (1352–1409), who wrote a preface for the text by royal command, focused on the interaction of geographic areas, environments, and bodily constitutions.

According to the preface of *Standard Prescriptions of Local Botanicals*, one of the most important principles of medicine is being in tune with multitude. It was significant to understand the manifold approaches to "diagnosing disease and prescribing medicinal drugs."[63] Every physician was supposed to manifest his own skill according to his own *ki*. Accordingly, there was no need to rely on a universal method in medicine. The author continued to argue,

In general, the method [*pŏp*] is different along the distance of 100 *ri*, and customs are different along the distance of 1,000 *ri*. The growth of wind, grass, and trees has its own appropriateness. The ways people eat, drink, enjoy, and desire also get accomplished in their own ways. Thus, the sages in the past tasted numerous herbs and followed the characteristics of every region, as the method of healing. Particularly, for our kingdom, heaven allots one region and allows us to inhabit the great East. Precious stuffs are hidden in the mountains and oceans, and the production of grass, trees and medicinal stuffs is possibly enough to support people's livelihoods. Consequently, those who treat people's diseases were also generally well prepared.[64]

This passage shows us what kind of medicine was envisioned to be the most appropriate for people in the Eastern Kingdom (Tongguk). Because the Eastern locale had developed its own customs, the potential for medical engagement already resided in its nature. Aspects of disease and medical treatments exhibited themselves differently in accordance with environmental variations. Corresponding to the characteristics of "our kingdom" (*aguk*), local products had already been used for curing "us." The ideal state of medical practice, however, was in contrast with the undesirable reality.

However, since bygone days, medicine has atrophied, and herbs are not collected in time. People ignore what is close [local] to them; instead, they seek what is far from them. Whenever people contract illness, they pursue Chinese medicinal drugs that are very difficult to get. Why does this happen? Isn't it just like pursuing a three-year-old wormwood for the treatment of disease that has been prolonged for seven years? As a result, medicinal drugs are unavailable, and the disease becomes serious to the extent that it cannot to be correctly treated.[65]

To rectify this dreadful situation, the author advocated medicinal drugs that fit the attributes of the soil, on which the formulas of "Eastern people" (*tongin*) would eventually demonstrate their efficacy. The resonant interrelationship among herb, soil, and body defies esoteric authority. Rather, a countryman who mastered the simple principle of medicine would properly treat "one disease by [adopting] a single herbal material."[66] A clear, profound command of native herbs, according to the author, would eventually allow people to seek solutions without going too far away, thereby easing their fear of dealing with diseases. As a way to compensate for the customary dependence on Chinese materia medica, the author argued for viewing medicine from a different perspective: not as universally defined principles, but as regionally elaborated ways of life. In this manner, the Eastern Kingdom was conjured up not as a marginalized backwater left behind by medical novelties but as a unique location for the creation of its own medicinal art.

LAYERS OF THE "LOCAL" IN THE EARLY CHOSŎN

The significance of the foregoing claim is beyond question. The court tried to disseminate *Standard Prescriptions of Local Botanicals* throughout the fifteenth century. In 1433, the text was sent to the southwestern Chŏlla and central-eastern Kangwŏn provinces.[67] In the same year, King Sejong ordered the magistrates of the northern provinces (such as P'yŏngan and Hamgyŏng) to pay special attention to managing local herbs. When new county offices were established along the northern border, *Standard Prescriptions of Local Botanicals*, in line with other Chinese medical texts, was sent for medical education there.[68] In 1454, the magistrate of the northern border area asked the court to send *hyangyak*; the court delivered both herbs and *Standard Prescriptions of Local Botanicals*. King Sŏngjong even suggested the translation of the summary of *Standard Prescriptions of Local Botanicals* into Korean vernacular to expand its circulation.[69]

In addition to court records, Hŏ Chun's *Precious Mirror of Eastern Medicine* employed *hyang* and *tang* to differentiate locally available herbs from their Chinese counterparts.[70] Hŏ Chun is the most renowned scholarly physician in Korean history, and his commitment to incorporating the term *hyang* suggests a continued concern for local botanicals in the elite composition of medicine.

Given this promotion of local botanicals, when we look more closely at the structure and sources of *Standard Prescriptions of Local Botanicals*, it becomes clear that this emphasis on indigenous herbs does not necessarily imply a radical departure from the dependence on Chinese medicine. Instead, the content, organization, and circulation of *Standard Prescriptions of Local Botanicals* reveals the early Chosŏn strategy of making the local meaningful under the influx of Chinese medical tradition.

A case in point lies with the sources on which the compilation relied. *Standard Prescriptions of Local Botanicals* elaborated on the content and structure of the source texts more fully than *Prescriptions of Local Botanicals for Emergency Use*, which had relied on the structure of *Collected*

Commentaries on Classical Materia Medica of the Heavenly Husbandman and *Illustrated Classic of Materia Medica.* The "Local Botanicals" part followed the main outline of *Materia Medica, Classified and Verified from the Classics and Histories.*[71] In summary, *Standard Prescriptions of Local Botanicals* continued to rely on Chinese sources as a major model and reference, updating the latest Chinese revision.

Even more telling is that only the last ten volumes out of eighty-five actually provided information on local botanicals, but the title, *Standard Prescriptions of Local Botanicals*, suggests that there would be a full elaboration of locally developed prescriptions. Following volume 76, which discussed the general principles of preparing and prescribing materia medica, the nine volumes from 77 to 85 elaborated on the traditional taxonomy of nature, classifying every entity under one of the ten categories of stones (*sŏk*), herbs (*ch'o*), woods (*mok*), people (*in*), beasts (*su*), birds (*kŭm*), bugs and scaly creatures (*ch'ungŏ*), fruits (*kwa*), grains (*kok*), and vegetables (*ch'ae*). The rationale of organization here emulates Chinese classics of materia medica. The grandiose compilation did not intend to rigorously catalog every herb found in the Chosŏn's territory. Instead, *Standard Prescriptions of Local Botanicals* aimed to exhibit the skill to reassemble the Chinese textual tradition of materia medica, which would display the Korean elites' mastery of the existing body of medical knowledge.

Like the Koryŏ's *Prescriptions of Local Botanicals for Emergency Use*, the names of herbs in *Standard Prescriptions of Local Botanicals* showed no clear rationale of nomenclature. Already well-known names, such as ginseng, did not need any distinction from Chinese ones. Whereas the Koryŏ's text simply explained ginseng's flavor, habitat, and collection method in about twenty-four Chinese characters, the Chosŏn's *Standard Prescriptions of Local Botanicals* added more content from Chinese sources, totaling about 380 Chinese characters.[72] Among all listed entities, only a few dozen names of herbs, fruits, grains, and vegetables were complemented with local names, *hyangmyŏng*. The *idu* names, introduced in the Koryŏ's text, disappeared; classical Chinese was the only authoritative language for nomenclature. Thus indigenous names of local herbs were not consistently rendered. Some of them changed during the late Koryŏ and early Chosŏn dynasties. The local names as labels in the Chosŏn's text were added complementarily to remind Korean learned audiences of the herbs' vernacular names.

It is also instructive to understand that *Standard Prescriptions of Local Botanicals* was hardly in popular demand. No record of further publication is found after the Chosŏn court republished the text in 1633. It survives today as a display in Handok Medico-Pharma Museum (Handok ŭiyak pangmulgwan) and the rare books division at the Kyujanggak Institute for Korean Studies, Seoul National University of Korea.[73] The dynasty's compilation of local botanicals was not frequently printed or readily consumed during the fifteenth and sixteenth centuries. Few individual scholars or physicians treated local botanicals as a theme of their texts on medicine. Not many bibliographies of medical formulas and texts that flourished between the eighteenth and early twentieth centuries referenced any of the grand compilations. Rather, the most widely circulated printed texts since the eighteenth century were classics of Chinese medicine, including the latest texts of materia medica written by a Qing scholar, eclectic formulae extracted from various medical literature, and a secret formula for healthy pregnancy and childbirth.[74]

Declining Interest in Local Botanicals?

The state project of promoting local botanicals was in decline by the early sixteenth century.[75] Kim Tu-jong ascribes it to the textual influx of medicine from Ming China. He points out that "prescriptions using the native medicinal drugs were hardly in use, and the collection of the native medicinal drugs were not encouraged, even left ignored. . . . The flood of Ming medicine that gradually flourished during the reign of King Sŏngjong has much to do with the decline of indigenous medicine."[76] Miki Sakae has related this preference for Ming medicine to the popularity of neo-Confucianism in Chosŏn Korea. Neo-Confucianism provided a milieu in which literati in Chosŏn could value textual inquiry into the medical classics and thereby aspire to the tradition of the Confucian sages. In line with this development, Song, Yuan, and Ming medical innovations became mainstream in Chosŏn Korea, marginalizing the indigenous knowledge of medicine.[77] The ideological establishment at court by the early sixteenth century saw a gradual decline of interest in local botanicals, which persisted in the following centuries.

The story of decline, however, leaves some more questions unanswered. Were the earlier complaints about the mismatch among herbs' morphology, names, and medical efficacy successfully resolved? How did the scholars and physicians after the sixteenth century respond to their customary use of herb names and related curative properties? Did they view the state's project of *hyangyak* in the early Chosŏn as completed? If not, what were the lasting problems associated with names of herbs in Korea?

Yi Sŏ-u (1633–1709), a well-known scholar-official, expressed disappointment over the continued misuse of and confusion around herb names in a poem. He ascribed the mischievous practice of medicine to the shallow understanding of herbs' names and nature.

> The study of materia medica done by the Divine Husbandman
>   [*Collected Commentaries on Classical Materia Medica of the
>   Heavenly Husbandman*, ca. 530–57] is outmoded and
>   seemingly doubtful
> Those herbs known to have medical efficacy again deceive people
> Barley is often confused with black soybeans
> How can we know about the bark of peonies without understanding
>   its flowers?
> Doctors misdiagnose even though I experience fracture in my arm
>   nine times
> Flavors are understood only after three times of tasting
> It's bitterly laughable! The names of the world are all false
> As false names give me a name, alas, who on earth am I?[78]

The author begins by underscoring the discrepancy between names in the Chinese classic of materia medica and the things he has experienced in his world. Incorrect herb names hardly ensured effective cures; furthermore, the accumulated confusions led to constant deception in people's minds. Given that the words in authorized texts did not correspond to things experienced in everyday life, the author casts doubt on the names that were given to posit his place in the world. The appropriate relationship between an herb and its name ultimately corresponds to the rightful positioning of a self in the world.

Yi Sŏ-u actually was not alone in criticizing the misuse of herb names. Although texts labeled as describing local botanicals largely vanished after the seventeenth century, efforts to clarify herb names continued as a part of the growing scholarly interest in "things and names" (*mulmyŏng*).[79] The growth of philological studies in late Ming and Qing China had an impact on the current of Korean "Practical Learning" (Sirhak), which employed pragmatic approaches to lexicology. During the eighteenth and nineteenth centuries, dozens of texts investigated the philological and phonological origins of words, exploring the lucid relationship between things and names.[80] Although these writings do not belong to the conventional genre of medicine, they pondered vocabularies related to herbs, vegetables, and trees.

It is only a small wonder that Yi Sŏ-u's poem was selected by Chŏng Yag-yong, the most renowned Practical Learning scholar, for insertion in his *Aŏn kakpi* (Realizing faults by rectifying words, 1819). Written at the high point of Chŏng Yag-yong's scholarship, the text discusses 450 terminologies categorized in two hundred articles. Among those articles, more than twenty discuss names of botanicals, such as cassia bark tree, maple tree, coptis root, and peppermint. Chŏng Yag-yong warned that the misuse of herb names, often found in clinical encounters, hampered medical efficacy. Not just the art of medicine was at stake; even more telling is that the foundation of knowledge would be lost by the distorted relationship between "name and actuality" (*myŏngsil*).[81]

To Chŏng Yag-yong, proper nomenclature establishes the moral ground of knowledge. He argues that "learning implies realizing," and "realizing hinges on differentiating the inappropriate [from the rightful]."[82] His moral ground of knowledge can be nurtured by clarifying the properness of words. Only by correcting the distorted meanings of each word would people reach the foundation of authentic knowledge. Understanding the origin of names, identifying the right characters with appropriate sounds, and knowing the paths toward the current modification of words became the ultimate foundation of trustful learning.

As depicted in Chŏng Yag-yong's criticism in its entirety, the partially dual and avoidably muddled use of names was intrinsic to Korean culture. In his epilogue to the study of names and things, Chŏng Yag-yong contrasted Eastern Kingdom Korea with Middle Kingdom China in their mastery of nomenclature.

In the Middle Kingdom, spoken word concurs with written word. If one expresses a thing from his mouth, then it becomes a written word, and vice versa. Accordingly, name and actuality coincide with each other and learned words communicate with vernaculars. However, this is not the case in our Eastern Kingdom. One simple example is sesame oil. In vernacular, it was called *ch'amgirŭm*, whereas in written words, it is rendered as *chinyu*. People only know *chinyu* is the learned expression, and are ignorant of other real names [*ponmyŏng*] of sesame oil, such as *hyangyu, homayu*, and *kŏsŭngyu*. More to the point, *naebok* in vernacular is called *muuch'ae*, shredded radish, yet people do not know this was misrepresented from *muhuch'ae*. Similarly, *songch'ae* is known as *paech'o*, napa cabbage in vernacular, yet this word is distorted from the original word, *paekch'ae*. Given these examples, it is enough to learn only one [name for a thing] in the Middle Kingdom, but in our country, even mastering three [words for a thing] is not enough.[83]

Chŏng Yag-yong argues that Koreans should pay special attention to clarifying the relationship between name and actuality because the spoken word did not fully correspond to the written word in Chosŏn Korea.

To be sure, naming the local botanicals in Eastern Kingdom Korea did not simply signal a shift toward an autonomous and self-sufficient sphere of medicine. Speculations on locally available botanicals, which continued even after the overall disappearance of *hyangyak* in the medical literature, indicated the different position of Eastern Kingdom Korea in giving names and ordering the natural world within its borders. Viewing the situation through the lens of local botanicals reveals the junctures where effective medical interventions, the ground of authentic knowledge, and the proper place of self in the world were closely connected but not always tuned harmoniously.

Conclusion

Comparatively viewed, the story of *hyangyak* in premodern Korea is consonant with analyses of indigenous herbs by Alix Cooper and Kapil Raj. Cooper persuasively demonstrates that early modern Europe's enthusiasm for exploring local flora was associated with the influx of exotic entities from other lands. While articulating the dichotomy of "indigenous versus exotic," professors of medicine in European college towns came to effectively manage their position as authoritative knowledge producers.[84] In a town or at a specific garden, the process of acknowledging indigenous flora and fauna often aims to speak to audiences outside of the investigated place. Cooper mentions that "the local knowledge set down by authors of local flora, while bearing clear evidence of their devotion to their adopted towns, was not primarily intended for the use or benefit of the townspeople themselves. It had another audience."[85]

Raj highlights that without a connection to an authoritative system, an isolated set of the indigenous fauna and flora hardly gains significance. Audiences and outside circulation networks actually define the success of a local botanical project. Nicolas L'Empereur (1660–1742), trained in medicine and a longtime resident of Orissa, India, completed his ambitious compilation, *Jardin de Lorixa*, a fourteen-volume folio on indigenous herbs. This project clearly relies on the knowledge and experience of local fakirs, merchants, and craftsmen. Contrary to L'Empereur's expectation, he was unable to gain support from his colleague in Paris, Antoine de Jussieu (1686–1758), a member of the Académie des Sciences and professor of botany at the Jardin du Roi. Jussieu refused to support L'Empereur's ambitious project, as he thought that exotic plants in their raw forms had nothing to do with European innovations in medicine. In other words, foreign botanicals that were not appropriately registered with European knowledge of medicine were useless.[86]

Similar features are found in the rise of *hyangyak*. Chosŏn Koreans obviously aimed to use locally available medical resources. This turn toward indigenous resources had to be balanced with the incessant desire for familiarity with the latest medical discourses and precious medicinal herbs from China. The emergence of local botanicals thus should be viewed as a strategy of managing the influx of medical knowledge and herbal entities in Korea, given the country's geopolitical position. This was one way to accommodate the more general and advanced learning at the periphery where the substance, agent, and framework of knowledge could not be exclusively under the Koryŏ or Chosŏn dynasty's control. Situated within the Sinocentric world order, Koreans did not necessitate the establishment of their own knowledge grid. In a sense, the accumulation of medical knowledge in Chosŏn Korea could hardly have been autonomous or independent. Simultaneously, however, it was necessary for Korean elites to impose the Chosŏn's positionality on the forms of medical knowledge, claiming indigenous differences and a geopolitical identity. Chinese medicine as a textual tradition was regarded as more authoritative, and adapting it to another locale calls for situated moderation. What has to be remembered is that the claim of local botanicals served as a medium to give meaning to a Korean intellectual project. It does not imply any rigid division between the Chinese and the local or the universal and the particular; rather, it alludes to a discerning way of relating the local to the more authoritative system of knowledge.

A comparison is enlightening when we examine the linguistic pluralism embedded in the European documentation of local botanicals. Cooper puts forth the "division of linguistic labor" to explain the different ways in which Latin and the vernacular (German) were both employed in cataloging local botanicals. Taking Christian Mentzel's (1622–1701) study about the local flora at Danzig as an example,[87] Cooper depicts the twofold nature of composing and consuming knowledge on indigenous herbs. In the catalog, Latin was used for nomenclature, thereby connecting each herb with the "universalistic world of naming and scholarly activity." On the other hand, German was used to describe specific places and habitats, such as names of mountains, rivers, valleys, mines, and nearby poorhouses, in connection with the examined botanicals. Different audiences reflected a kind of division of linguistic consumption. Whereas the university community embraced Latin and its embedded atmosphere of cosmopolitanism, readers from town councilors to local noble families responded to the unique nature of botanicals "in the contexts of particularity and of 'popular' associations."[88] In this regard, the parallel uses of Latin and German exemplify harmonious cooperation rather than conflict or competition for authority.

Texts about local botanicals in premodern Korea did not display the same degree of mutual collaboration. Although the state project of local botanicals paralleled the invention of the Korean alphabet (1446), no texts of local botanicals were written in vernacular Korean. Chinese characters were used to express the popular pronunciation of herb names. As long as Chosŏn Korea willingly embraced the civilizational ideal of Sinocentrism, the deviation from classical Chinese highlighted its sense of inferiority.

William Crossgrove's analysis, therefore, sheds light here. As he puts it, "Vernacularization refers to the transposition of texts from a high-status language, usually Latin, into a vernacular language that typically has lower prestige as a written language." Accordingly, the hierarchical relationship assumes that the indigenous sphere of knowledge is in opposition to "something more learned, more conscious, more prestigious."[89] Inasmuch as the invention of the Korean alphabet was despised by some Korean elites, the composition of local botanicals in the vernacular was hardly valued. Compared to the European "division of linguistic labor," the Korean insertion of the vernacular, even though in Chinese character transliteration, was complementary and adjusted to the existing hierarchy.

In addition to *hyangyak*, "Eastern medicine," or *tongŭi*, has identified the geocultural distinction between (Chosŏn) Korea and its Chinese and Western counterparts. The next chapter traces the term's origins in the seventeenth century, simultaneously exploring its modification under the growing sense of nationalism and colonialism in the late nineteenth and early twentieth centuries.

CHAPTER TWO

Eastern Medicine, or *Tongŭi*

Imagining a Place for Medical Innovation

Is it appropriate to write a book, if sages in the past have already completed what is worth being written?" Hwang To-yŏn, a renowned scholarly physician, began his voluminous medical anthology, *Ŭijong sonik* (Foundations of medicine revised, 1868), by questioning the legitimacy of book writing.[1] The sages had already disclosed the truth about life and the body, and the past millennia had only witnessed increasing confusions and misconceptions. What, then, is the value of producing textual knowledge of medicine in the present? The following quote hints at Hwang To-yŏn's answer:

> Since the Yellow Emperor and his retainer Qibo [since medicine has begun], among the books of the three sages, what survived intact down to the Qin dynasty [of China] were only the books of medicine. What is called medicine is a matter of timeliness. What was appropriate for the past [for the former times] is only up to that time. At all times and in all countries under heaven, the principle [of medicine] is one. However, people's diseases are extremely different. According to [different types of] people's disease, medicinal drugs are also different in their use. Furthermore, the climate [natural features] of the region, the administration of the south and the north are quite different from each other. The human inborn constitution and the pharmaceutical preparation of the warm and the cold all follow those differences respectively. In addition, there are changes between past and present, which are not similar at all.[2]

Here Hwang To-yŏn primarily recognizes a physician's authority in relation to the medical sages of the past. Given the sacred origin, the trustworthy ground of medical learning cannot be equivocal. He clearly states that the principles of medicine are one in all times and in all countries under heaven. Elsewhere in the preface he criticizes those who "ignore the prescriptions that sages have created and preserved." Simultaneously, however, Hwang To-yŏn alludes to the limitation of the past in grasping the entirety of illness variation in the present. Evolving diseases and their various manifestations exist in all stages of human life. Accordingly, along with regional and temporal variations, differences of human physiology are seen to affect therapeutic solutions.

Given the tension between endlessly transforming diseases and the supposedly constant principle of medicine, Hwang To-yŏn argues that the most important lessons for physicians involve timeliness, or appropriate intervention: "What is called medicine is a matter of timeliness." According to him, being a good doctor implies being keenly aware of a present problem in the light of past knowledge. What is clear, yet subtle, in his discussion is the negotiation between the wish to overcome the past and the desire to rely on the past for guidance. For Hwang To-yŏn, the true authority of medicine was to be illuminated not by blindly following past tradition nor by radically deviating from it, but by appropriately situating one's problem within the medical tradition that seamlessly flowed down to the present.

Hwang To-yŏn's call for balancing diachronic divergence, as the virtue of medicine, helps us consider the evolution of "Eastern medicine" in the textual tradition of Korean medicine. Inasmuch as the gap between past and present was recognized in articulating the art of medicine, the regional differences in environment and administration would affect the management of health and illness. In addition to labeling the South and North, as seen in Hwang To-yŏn's preface, labeling the East is part of the Korean production of medical texts. What is the role of combining the East with medicine? As is discussed in the introduction of this book, no term better represents the indigenous attributes of Korean medicine than "Eastern medicine." The place-specific distinction has been unanimously employed by North and South Koreas in authorizing traditional medicine in the twentieth century. Without a doubt, the East mapped on medical heritage has proudly signified the demarcated realm of indigenous medicine in contrast to its Chinese and Western counterparts.

Beyond its popular recognition, however, we know little about the patterns of using "the East" in the elite compilation of medicine. What are the origins of the term? What are major features of Eastern medicine in terms of textual components, therapeutic solutions, or the establishment of medical lineages? Last but not least, in what ways has the geographical sensitivity of Korean medicine articulated the boundaries or connections among people, texts, and authority? What light, if any, can Hwang To-yŏn's balanced approach to the diachronic tension shed on the geographic distinction expressed in Korean medical texts?

Eastern Medicine and the *Treatise on Cold-Damage Disorders*

For an overview of Eastern medicine, this chapter begins by examining the two best-known Korean medical texts, Hŏ Chun's *Precious Mirror of Eastern Medicine* and Yi Che-ma's (1838–1900) *Tongŭi suse powŏn* (Longevity and life preservation in Eastern medicine, 1894). Of all the medical texts published in Korea before 1900, only these two include Eastern medicine in their titles, exhibiting a sense of regional or cultural distinction. The history of Eastern medicine thus cannot be pursued without delving into those two texts. (Dis)continuity between the two compilations reveals different editorial goals and designs, changing motivations, and (dis)similar patterns of therapeutic priorities. After analyzing the elements of Eastern medicine manifested in those two texts, this chapter examines the legacy of Hŏ Chun and Yi Che-ma during the rule of the Japanese colonial government (1910–45), elucidating how a growing sense of nationalism and an urgent longing for medical modernization reshaped the organization and meaning of Eastern medicine.[3]

A key to historicizing Eastern medicine lies in the Korean composition of Zhang Ji's (150–219) *Treatise on Cold-Damage Disorders*, which is crucial in understanding the language, editorial framework, and therapeutic priorities of Eastern medicine over time. First and foremost, the history of the *Treatise on Cold-Damage Disorders* testifies to the significance of textual (re)assemblage as a way of producing medical knowledge. Zhang Ji's original writings were lost; at intervals, numerous unsuccessful attempts have been made to fully recover the ancient text. The idea of the authentic *Treatise on Cold-Damage Disorders* in its original form, with complete sentences, organization, and prescriptions, may diminish our regard for the multiple versions of this lost classic published over time.[4] The different modes of compiling, editing, and annotating this work teach us not just to measure the degree of fabrication or misuse but to understand the extent to which each version reflected its author's desire to improve medical knowledge and practice. The *Treatise on Cold-Damage Disorders* in the elite compilation of Korean medicine indicates the warp and woof of Eastern medicine as a textual production.

Second, the *Treatise on Cold-Damage Disorders* is foundational, as it introduces "six-channel pattern identification" (*yukkyŏng*) to categorize diseases. No single explanation has clarified the precise meaning of six-channel pattern identification. It may denote meridians (*kyŏngnak*), a path of flowing *ki*, or topological differentiation of the body.[5] With a flexible mode of interpretation, the six categories in the learned medicine of East Asia have served as classificatory nomenclature through which the manifestations of ever-changing illness are conceptualized. Through the six layers of differentiation, the progression of disease was related to various degrees of cold or heat and the unbalanced state of depletion and repletion of the body, which eventually led to relevant prescriptions. Without a complete philosophical articulation, the *Treatise on Cold-Damage Disorders* has served as the clinical guidance for traditional medicine in East Asia. The issue, then, lies in the degree of Korean accommodation or modification of the major terminologies, conceptual framework, and prescriptions of the *Treatise on Cold-Damage Disorders*. Did Korean physicians fully embrace or challenge the authoritative language and therapeutic solutions of the *Treatise on Cold-Damage Disorders* in the formation of major texts labeled as Eastern medicine?

Finally, the *Treatise on Cold-Damage Disorders* exemplifies how a revered medical text has been reconfigured by later generations in an effort to cope with new diseases and maintain authentic medical lineages. The idea of cold damage—the attack of an external pathogen—covers illnesses that range from the common cold to all diseases involving fevers and other infectious diseases, such as acute febrile epidemics.[6] Marta Hanson, in her book on warm disease (*wenbing*), persuasively demonstrates that the indigenous Chinese disease concept was originally understood under the orthodox rationale of cold damage in the *Treatise on Cold-Damage Disorders*, then emerged as a separate disease category with its own distinct etiology and therapeutics around 1642.[7] The severe epidemics in the late Ming dynasty period only testified to the ineffective prescriptions of the *Treatise on Cold-Damage Disorders*. Shifting away from the cold-damage framework of the treatise, the scholars and doctors in the southern Jiangnan region specified a pestilential or deviant *qi* as the major cause of most severe warm epidemics and accordingly required a geographically articulated rationale of diagnosis and therapies. The growing number of publications about warm disease beginning in late Ming was associated with the rising local initiative based on both a family lineage and the regional social networks of the medical community. The Chinese appropriation of the *Treatise on Cold-Damage Disorders* over time, in line with the southern rationalization of epidemics, cannot be applied mechanically to Korean history. However, the (re)composition of the *Treatise on Cold-Damage Disorders* by Korean elites discloses the connections and divergence among textual authority, therapeutic principles, and the place-specific reasoning in medicine, thereby helping us further analyze Eastern medicine as a textual production.

I argue that the Korean understanding of the *Treatise on Cold-Damage Disorders* illuminates one of the building blocks of Eastern medicine. The passages of text in Hŏ Chun's *Precious Mirror of Eastern Medicine* exemplify a set of editorial techniques that enabled him to argue for the positionality of Eastern medicine. Yi Che-ma's interpretation of the *Treatise on Cold-Damage Disorders* illustrates his ambitious medical synthesis, which begets major terminologies of the currently known "Four Constitutions medicine" (Sasang ŭihak). Finally, the Korean comprehension of the *Treatise on Cold-Damage Disorders* in the early twentieth century hints at novel patterns of constructing Eastern medicine under the growing anxiety about Western medicine. As Yamada Keiji aptly puts it, the *Treatise on Cold-Damage Disorders* in history serves as a "landscape."[8] Taking the text as a window, we can look over each historical juncture as it (re)assembles and (re)interprets a medical classic on its own terms. Analyzing the shifting montage of the treatise in Korea's major medical texts is indispensable for further understanding the textual, therapeutic, and cultural connotations of Eastern medicine.

The *Treatise on Cold-Damage Disorders* in Korea between the Seventeenth and Nineteenth Centuries

COLD-DAMAGE DISORDERS IN HŎ CHUN'S COMPILATION

General Organization. There is no record detailing exactly when the *Treatise on Cold-Damage Disorders* was imported to Korea. A few scattered documents testify that medical teachings about cold-damage disorders were available in Korea as early as the sixth century CE. During the

Koryŏ dynasty, the teachings about cold-damage disorders were published as part of a state project.[9] The court record states that *Zhang Zhongjing wuzhanglun* (Discourses on Zhang Zhongjing's five inner organs) was published in 1059, along with dozens of Chinese medical texts.[10] In 1092, the Chinese court asked the Korean envoy whether the Koryŏ dynasty held woodblock copies of a series of texts that were missing in China. Among them were a dozen medical works, including one comprising fifteen volumes that was titled *Zhang Zhongjing fang* (Formulary of Zhang Zhongjing).[11]

During the Chosŏn dynasty, the studies about cold-damage disorders became essential for educating court doctors.[12] More important, the *Classified Compilation of Medical Prescriptions*, which aimed to express the newly founded Chosŏn dynasty's cultural competence and pride, documented the full contents of the *Treatise on Cold-Damage Disorders*. The *Classified Compilation of Medical Prescriptions* ambitiously sought to pull together every known piece of medical literature from the Han, Tang, Song, and Yuan dynasties and sum them up in 266 volumes (*kwŏn*). In this grand state project to publish a medical encyclopedia, explanations and prescriptions for cold-damage disorders are listed under a few subcategories.[13]

Not until Hŏ Chun, however, was the *Treatise on Cold-Damage Disorders* incorporated into an individual author's medical reasoning and editorial design. In putting forward his own principles of organization, selection of terms, and interpretation of previous scholarship, Hŏ Chun represents a departure from previous Korean organization of cold-damage disorders.[14] Favoring the latest medical writings from Yuan and Ming China, Hŏ Chun examined more than two hundred existing medical texts. After selecting more than 2,000 symptoms, 1,400 medicinal substances, and 4,747 remedies, he organized them into five major sections (*p'yŏn*) and twenty-five volumes (*kwŏn*): "Interior Landscape" (*Naegyŏng*), "External Forms" (*Oehyŏng*), "Miscellaneous Diseases" (*Chappyŏng*), "Decoctions" (*T'angaek*), and "Acupuncture and Moxibustion" (*Ch'imgu*).[15] Discussion about cold-damage disorders fell mostly into the "Miscellaneous Diseases" section, under the category of "cold" (*han*). As one of the five major sections of the *Precious Mirror of Eastern Medicine*, "Miscellaneous Diseases" comprises eleven volumes, which represents the greater part of the text, and details the causes, manifestations, and treatments of various diseases.

Hŏ Chun's elaboration of miscellaneous diseases begins by explaining teachings on the "environmental and bodily circles of *ki*" (*ch'ŏnji un'gi*), suggesting that proper treatment implies more than just examining an individual body. Once the environmental conditions of diseases have been explained, Hŏ Chun goes on to detail the principles of "diagnosis" (*simbyŏng*), "differentiating manifestations" (*pyŏnjŭng*), "pulse taking" (*chinmaek*), "consuming medicinal decoctions" (*yongyak*), and the three methods of "vomiting" (*t'o*), "sweating" (*han*), and "purging" (*ha*). There follows an account of six etiological categories: "wind" (*p'ung*), "cold" (*han*), "heat" (*sŏ*), "humidity" (*sŭp*), "dryness" (*cho*), and "fire" (*hwa*).[16] Here, "cold" provides an organizational scheme under which Hŏ Chun elaborates in detail the range of symptoms caused by various exogenous factors. His composition begins with general principles and then extends to more concrete illness categories.

Cold Damage under "Miscellaneous Diseases." Hŏ Chun's overarching category of "Miscellaneous Diseases" is quite distinct from the way his contemporary Chinese scholarly doctors organized cold-damage disorders. For instance, Li Chan's (active 1573–1619) *Yixue rumen* (Introduction to medicine, ca. 1575), which Hŏ Chun substantially refers to for his explanation of cold-damage disorders, separates the category of "cold damage" (*sanghan*) from "miscellaneous diseases" (*chappyŏng*), making this a main organizing thread. Unlike Li Chan, Hŏ Chun puts "cold damage" under the category of "Miscellaneous Diseases," which implies his emphasis on the general rationale in treating all manifestations of illnesses. He pays less attention to the particular nature of externally contracted diseases, subjugating them to the general principle of treating various diseases.

In "Miscellaneous Diseases," under the category of "cold," Hŏ Chun introduces passages from the *Treatise on Cold-Damage Disorders*. Emulating Zhang Ji's six-channel pattern identification as an organizational principle, greater yang (*t'aeyang*), yang brightness (*yangmyŏng*), lesser yang (*soyang*), greater ŭm (*t'aeŭm*), lesser ŭm (*soŭm*), and reverting ŭm (*kwŏrŭm*) provide the entry points of different symptoms. To be sure, Hŏ Chun articulates more information about cold damage. For instance, he introduces a range of general principles before elaborating the details of the six-channel pattern identification. These teachings expand on general attributes of cold damage, such as "cold-damage disorders contracted in winter," "cold-damage disorders are serious illnesses," "method of pulse taking," "dates of recovery or death," and so on. In addition, after discussing the six-channel pattern identification, Hŏ Chun adds a range of symptoms that were considered manifestations of cold damage. Throughout this rigorous exposition, however, the six-channel pattern identification stands out as a major organizing principle. As a significant editorial thread, each channel is elaborated with Hŏ Chun's summary of noticeable symptoms, warnings, and prescriptions.

The "cold" section, like the other parts of the *Precious Mirror of Eastern Medicine*, points to a series of Hŏ Chun's sources. In addition to the *Treatise on Cold-Damage Disorders*, a range of Chinese texts, such as *Introduction to Medicine, Huoren shu* (A book of saving life, 1088), *Gujin yijian* (A mirror of medicine of all times, ca. 1589), and *Yixue zhengzhuan* (Orthodox transmission of medicine, 1515), provided the language and rationale for his synthesis. Hŏ Chun's sources for the "cold" section overlapped with his general preference for the latest medical literature from China. Out of his ten most frequently referenced texts, seven come from the authors of Jin, Yuan, and Ming China.[17] More than 90 percent of his sources are traceable because he indicates the reference by the name of a text or an author with each unit of composition.[18] Although he follows the guideline of "writing what is already written by tradition without adding personal opinions" (*suribujak*), Hŏ Chun implicitly inserts his point of view by intentionally selecting, omitting, and rephrasing the original Chinese literature.

Language of Dualism as a Structural Theme. To further understand Hŏ Chun's craft of textual composition, it is necessary to focus on his explanation concerning the greater yang channel, one of the six-channel pattern identifications. This is one of the most well-known parts of the *Treatise on Cold-Damage Disorders*, and it has wide application to the first stage of externally contracted diseases. To understand Hŏ Chun's (re)assemblage, let us take a look at the first few lines of the greater yang entry. The *Treatise on Cold-Damage Disorders* begins with the pulse diagnosis and salient symptoms. Here are the first three lines of this entry from Zhang Ji's original:

In disease of the greater yang, the pulse is floating, the head and nape are stiff and painful, and [there is] aversion to cold.

When in greater yang disease [there is] heat effusion, sweating, aversion to wind, and a pulse that is moderate, it is called wind strike.

Greater yang disease, whether heat has effused or not, as long as there is aversion to cold, with generalized pain, retching counter flow, and yin and yang [pulses] both tight, is called cold damage.[19]

These lines can be compared with those of Hŏ Chun, who explained the greater yang channel by combining Li Chan's *Introduction to Medicine* with Zhang Ji's original. As Hŏ Chun puts it,

> If diseases reside in the greater yang channel rooted in the bladder, the head is painful, the back is stiff. The small intestine becomes the sign, and with the heart, [the small intestine] presents the exterior-interior relationship, and there is heat effusion. Ephedra Decoction and Cinnamon Twig Decoction are good in winter and Nine-Herb Decoction with Notopterygium for other seasons.
>
> For the greater yang diseases, regard skin as the exterior and bladder as the interior. If heat resides in the skin, then the head is painful, the nape is stiff. This should be treated with Ephedra Decoction, Cinnamon Twig Decoction, and Nine-Herb Decoction with Notopterygium. If heat resides in the bladder, the patient feels thirsty, and the urine is red, then Five Ingredient Powder with Poria should be given. (*Introduction to Medicine*)
>
> [If there is] heat effusion, aversion to cold, and the pulse is floating, these belong to the exterior, and this is the pattern of the greater yang. (*Treatise on Cold-Damage Disorders*)[20]

What is important here is Hŏ Chun's selection of the original literature, Li Chan's *Introduction to Medicine*, which used terms of dual manifestation. Hŏ Chun prioritizes understanding the paired relationship between outer manifestation and internal organs, or reading the surface sign in connection with its inner phenomena, in his understanding of diagnosis and treatment. Consequently, his sentence selection frequently reveals double layers: "surface sign" (*p'yo*) and "root" (*pon*) or the "exterior" (*p'yo*) and "interior" (*ri*). This applies not only to the passage above on the greater yang channel but also to the entries for the other five channel patterns, which exhibit similar wording. The dual nature of disease manifestation is a central idea for Hŏ Chun. Before the greater yang section, he offers the following sentence from the *Introduction to Medicine* as a guideline: "The channels become the sign, and internal organs become the root. For instance, the greater yang channel becomes the sign and the bladder the root. The other channels are the same. (*Introduction to Medicine*)."[21] According to him, it was not enough to simply read the patterns of symptoms and match them to relevant prescriptions: deeper understanding of the cold-damage disorders lies in the balancing assessment of any resonance between the surface sign and interior root or the outer manifestation and internal organ functions.[22]

Hŏ Chun's emphasis on dual relationships parallels his design for the structure of the entire *Precious Mirror of Eastern Medicine*. He suggests that the "Internal Landscape" and "External Form" sections should be considered essential guides to probing the body for signs of health or illness. Understanding the "Inner Landscape" section is important, as it shows his theory of the "essence" (*chŏng*), "vital energy" (*ki*), and "spirit" (*sin*) as significant attributes of the body. The "External Form" section, which discusses physical body parts such as the head, face, back, breast, and abdominal region, parallels the inner functions.

Hŏ Chun's organizing principles partly reflect his intention to position medicine within a philosophical framework. In the preface, he argues that his compilation of medical texts agrees with the rationale of Daoist classics like *Huangting jing* (Scripture of the yellow court, ca. 3rd century). *Scripture of the Yellow Court* deals with the "internal landscape," and certainly this is echoed in Hŏ Chun's discussions of the "contours of interior and exterior boundaries and shapes" (*naeoe kyŏngsang chi to*).[23] At the same time, he modestly acknowledges the limitations of medical understanding through comparison with the totality of the Daoist perspective: "Daoist teaching takes pure essence and nurturing life as the foundation of life, and medical learning takes medicine and acupuncture as the foundation of treatment. Therefore, Daoist teaching deals with the entire mind and body carefully whereas medical teaching deals with only a part."[24] Conforming to the synthetic principle of "Three teachings in one" (*samgyo habil*), Hŏ Chun prioritizes teaching and practicing "nourishing life or health preservation" (*yangsaeng*), which was also prevalent in the late Ming period.[25] In short, his language of duality conforms to his organizing framework, reflecting his desire to address diseases and health through a holistic understanding of human beings. Hŏ Chun aimed to go beyond a simple and exhaustive cataloging of diseases and therapeutic solutions. The terminology and organizational framework of his *Precious Mirror of Eastern Medicine* reflects his desire to achieve intellectual coherence in describing manifestations of cold-damage disorders.

As the foregoing example suggests, the passages of the *Treatise on Cold-Damage Disorders* in the *Precious Mirror of Eastern Medicine* partly reflect Hŏ Chun's general approach to composing a medical anthology. In response to the flow of medical literature until the Ming dynasty, Hŏ Chun incorporates the original Chinese teachings within his own key terminologies and structural design. This is evident in the "internal and external" or "sign and root" distinction that informs his organizing framework, selection of references, and choice of words. Even though these ideas had already and fre-quently been expressed in Daoist classics and Ming medical texts,[26] Hŏ Chun wanted to give philosophical coherence in his synthesis, moving a step away from previous methods of compiling medical texts in Korea. As Fabien Simonis persuasively puts it, the "coherent amalgamation" was achieved by "deliberate syncretism" or "eclecticism by design," which char-acterizes the *zhezhong* style of intertextuality in Ming medical texts.[27] Only sophisticated scholars could selectively reorganize passages from canonical texts, navigating complicated overlaps and divergences. The balanced as-sessment of textual heritage reflects maturity, not amateurism.

Hŏ Chun's confidence in reorchestrating the existing repertoire of Chinese medical texts solidified his organizational coherence; accord-ingly, he labels his encyclopedic compilation with the distinction of the East. He argues that since Chinese physicians such as Li Gao and Zhu Zhenheng were taken to represent "Northern" and "Southern" medicine, respectively, his own synthesis, with the phrase "Eastern medicine," deserved a further geocultural distinction. Although "our kingdom is remotely situated in the East," he states, "the way of pursuing medicine has never been stopped here."[28] Hŏ Chun's Eastern medicine therefore re-flects his will to identify his position in the wider world of medical thinking and practice. Although located on the margin of the Sinocen-tric world, according to Hŏ Chun, the Chosŏn kingdom had confidently elaborated its own perspective of health and illness.

HŎ CHUN'S WORK AS
ORGANIZATIONAL FRAMEWORK

After Hŏ Chun, major publications tended to follow his outline in com-piling a range of symptoms described as "cold damage." For instance, court physician Kang Myŏng-gil (1737–1801) composed *Chejung sinp'yŏn* (New compilation for benefiting people, 1799) in eight volumes (*kwŏn*) and five books (*ch'aek*) and aimed to emulate the *Precious Mirror of East-ern Medicine*, but omitted what he thought to be superfluous content and added the latest prescriptions.[29] Sponsored by King Chŏngjo, Kang Myŏng-gil intended to produce a practice-oriented text while relying on and going beyond Hŏ Chun's philosophical stance. In the preface, Yi Pyŏng-mo (1742–1806) asserts that Kang Myŏng-gil's compilation "takes out complicated sentences and complements missing parts aiming at be-ing fully selective and precise to produce a well-organized text.... Even people in a remote place may benefit from this book and easily access medicine according to their own symptoms."[30]

In terms of organization, Kang Myŏng-gil places the six etiological categories—wind, cold, heat, humidity, dryness, and fire—in the first volume; in the second, third, and fourth volumes, he goes on to summarize Hŏ Chun's interior landscape and external forms. Kang Myŏng-gil's choice to begin with the six etiological categories seems similar to Li Chan's decision regarding the first part of his *Introduction to Medicine*. Under the "cold" section, Kang Myŏng-gil organizes a range of "cold damage"–labeled symptoms and prescriptions by prioritizing the six-channel pattern identifications, which he views as categories of illness, and then matches these to relevant prescriptions.[31]

Compared to Hŏ Chun, Kang Myŏng-gil's terminology, layout, and selected prescriptions are considerably more concise and straightforward. The section on greater yang, for instance, states symptoms directly without tracing the underlying dynamics: "The head is painful, the body has heat, the back is stiff. Without sweating, [there is] aversion to cold. If the pulse of *ch'ŏk* and *ch'on* is floating and tense, this is cold damage (*sanghan*); if the pulse of *ch'ŏk* and *ch'on* is floating and relaxing, [it is] wind damage (*sangp'ung*)." After this brief statement, he introduces only one prescription from Hŏ Chun's original, Nine-Herb Decoction with Notopterygium (*kumi kangwalt'ang*), believing that the decoction most evidently works for the range of symptoms that might fall into the greater yang category. The same pattern continues in the remaining five categories. Kang Myŏng-gil eliminates Hŏ Chun's theoretical terminology of dualism; he does not elaborate Hŏ Chun's framework of double layers ("surface sign" and "root") or assign any central importance to the interplay between internal organs and external manifestation. Kang Myŏng-gil also eliminates philosophical discussion of the "environmental and bodily circles of *ki*" and other general principles that Hŏ Chun included in his "cold" section, such as "diagnosis" and "differentiating patterns." In his list of references, he includes twenty-one texts, mostly from Yuan and Ming China. Yet for composing the section on cold-damage disorders, he refers heavily to Hŏ Chun's *Precious Mirror of Eastern Medicine*.[32] Selection and summarization can be properly rendered through trained eyes; in this regard, the section on cold-damage disorders in Kang Myŏng-gil's abridged compilation reflects his confidence in mastering Hŏ Chun's anthology.

In his *Ŭibang hwalt'u* (Essential prescriptions, 1869), Hwang To-yŏn presents an editorial aim similar to Kang Myŏng-gil's, although he organizes his text for greater convenience as a clinical guide. In composing the section on cold-damage disorders, Hwang To-yŏn also puts six environmental categories in the first volume and deploys the six-channel pattern identification as quasi-nosological categories. He also adopts the three ways of classifying medicinals, which originated from Tao Hongjing's *Collected Commentaries on Classical Materia Medica of the Heavenly Husbandman*. All prescriptions were organized according to the replenishing (upper), harmonizing (middle), and attacking (lower) natures. In this way, a series of symptoms that might belong in the greater yang category could be treated with three different kinds of prescriptions.

In summary, Korean elite publications after Hŏ Chun included passages about the cold-damage disorders according to their own editorial preferences and clinical priorities. In their respective approaches to cold-damage disorders, Kang Myŏng-gil and Hwang To-yŏn do not deviate radically from Hŏ Chun's organizational principles, working with due adherence to Hŏ Chun's editorial framework, his prescriptions, and his selection of original Chinese references. Even so, the textual component of Hŏ Chun's Eastern medicine survived with a degree of subtraction and reorganization.

A THERAPEUTIC GUIDELINE
FROM HŎ CHUN?

Does Hŏ Chun's enduring framework for composing the section on cold-damage disorders indicate any equivalent pattern in clinical encounters? How were illnesses understood as cold damage actually treated? A couple of medical cases in Chosŏn Korea hint at a trend in treating cold damage.[33] For instance, one of Ŭn Su-ryong's (1818–97) eleven extant cases demonstrates his principle for curing externally contracted "cold ache" (*hant'ong*).[34] In this case, Ŭn Su-ryong describes how he successfully treated his friend's daughter-in-law, wife, and maid. Following the sudden death of his friend's mother, the family was suffering from emotional exhaustion and physical hardship during and after the mourning period. The symptoms common to the three women included feelings of a sudden chill and collapse due to fever and headache. Ŭn Su-ryong prescribed milk vetch root as a primary ingredient, adding ginseng and white atractylis (*paeksul*) as secondary ones, and then complemented this prescription with small amounts of bupleurum (*siho*), dried orange peel (*chinp'i*), *Saposhnikovia divaricata* (*pangp'ung*), and angelica root (*kanghwal*). He also suggested that the patient's body be warmed up without sweating. After a day or so, according to him, all the women had recovered. Ŭn Su-ryong argued that the efficacy of his cure had to do with his principle for dealing with cold-damage disorders. He describes the case as follows:

> It was winter and [these people] did not rest fully after overwork. Thus, the *ki* of vicious cold (*hansa*) made the most of the emptiness, then primarily invaded the first meridian of greater yang. Accordingly, the vicious *ki* from outside are fighting against the appropriate *ki*, which aims to block the advancement of the vicious *ki*. It is needless to consider any other negligible symptoms. I only prioritize to replenish the appropriate *ki*. If this appropriate *ki* flourishes, the vicious *ki* will fade away by itself even without attacking it with additional prescriptions. If the appropriate *ki* protects, how would the vicious *ki* go deeper through the manifestations? That does not make sense at all. To sum up, this time as well, I stick to the lesson of "Do not purge although [the patient] is externally contracted." In other words, I did not violate the rule that even though it is caused externally, you should not use the medicine of purgation. If I ignore the principle and prescribe the medicine of "attacking the poison and governing with discharge," then the patient's symptoms of emptiness become much worse, and I am not sure whether the patients would live or die.[35]

A case included in Yi Su-gi's (1664–?) *Yŏksi manp'il* (Miscellaneous jottings on medical experiences and tests, 1734) reveals a similar inclination toward the principle of harmonizing and replenishing. Yi Su-gi criticizes a treatment that aimed to attack the "cold," as the overuse of cold medicine worsened a young man's "cold-damage manifestation" (*sanghanjŭng*). He reports that "the son of Yi Saeng got married at the age of eighteen. On the tenth lunar month, after dozens of days of his consummation, he contracted the manifestation of cold damage. . . . After diagnosis, I told with surprise that 'this is not the manifestation of yang or real fever (*siryŏl*). If you keep overtaking cold medicine, it will certainly worsen the patient's condition.'" Yi Su-gi prescribed the Ophiopogonis Decoction (*maengmundongt'ang*), adding two *ton* of ginseng.[36]

Ŭn Su-ryong agrees with Yi Su-gi that a body depleted by sexual exhaustion is particularly susceptible to cold damage. In another case, Ŭn Su-ryong reports that a recently married young man was suffering from "symptoms of ŭm" (*ŭmjŭng*). The man had married at the age of twenty-one but also had extramarital relationships. He complained of frequent chills, fever, and aches in his limbs and joints, and family members used "medicine that discharged internal fever by sweating" (*palp'yo hwahae chi che*). The patient did not show any improvement, and in fact became worse after more than twenty days in this condition. Ŭn Su-ryong reasoned that the young man's exhausted body, the depletion of appropriate *ki*, had caused the onset of vicious cold, but fortunately, the cold damage had not reached the deeper layers. What Ŭn Su-ryong observed, then, was the earlier stage of *sanghan*, "manifestation of yang" (*yangjŭng*), so he applied the principle of replenishing, prescribing medicine that protects the original *ki*.[37] Eventually, the patient was cured.

Clearly, Yi Su-gi's and Ŭn Su-ryong's cases do not represent the entirety of cold-damage treatments in late Chosŏn Korea. However, their cases indicate a persistent assumption in therapeutic approaches: that even the destructive power of an external attack cannot penetrate the body as long as that body is not depleted. It is interesting to note that this hesitance in using the cold and discharging medicine was also found among Korean practitioners in the 1930s, who criticized the Japanese *shanghan* specialists' bias toward aggressive medicine. It is no exaggeration to say that in general, this doubt about purgative medicine and a belief in the body's protecting and rejuvenating potential shape the general ground for Korean practitioners in treating illnesses labeled as "cold damage."

In the following section, I discuss how Hŏ Chun's strategy in assembling the *Treatise on Cold-Damage Disorders* was (dis)continued by Yi Che-ma, who is also celebrated by Koreans as one of the most important figures of the indigenous medical heritage.

The *Treatise on Cold-Damage Disorders* in Nineteenth-Century Korea

YI CHE-MA'S EASTERN MEDICINE: DIVERGENCE AND CONTINUITY

As a nineteenth-century Korean physician-intellectual, Yi Che-ma hardly enjoyed the feeling of being privileged. Not much is known about his personal life, but a few studies point out that he was born as an illegitimate son (*sŏŏl*) to a *yangban* family and left his hometown at the age of thirteen, after his father and grandfather passed away. After wandering around Korea and Manchuria, he began his career as a military officer at the age of thirty-nine and was appointed a county magistrate (*kunsu*). However, he does not seem to have enjoyed the highest position he gained throughout his life. He returned to his hometown right after the appointment and spent the rest of his life practicing and writing about medicine.[38]

Compared with Hŏ Chun, in his own assemblage about the cold-damage disorders, Yi Che-ma voiced his disagreement with Zhang Ji more explicitly. First and foremost, Yi Che-ma discarded the six-channel pattern identification and suggested his own classificatory principle centered on human differences. According to him, Zhang Ji focused on "disease patterns" (*pyŏngjŭng*), whereas Yi Che-ma prioritizes "people" (*inmul*). Overlaps are found in the terminologies they used, but Yi Che-ma advises readers not to equate his classificatory categories with those of Zhang Ji.[39] Four types of people are identified with the categories of lesser ŭm, lesser yang, greater ŭm, and greater yang. Each category reflects the dissimilar nature of human physiology, the strength and weakness of the visceral system, the uneven senses of morality, and emotional extremes. These categories are rigid and deterministic in the sense that they hardly change over time, although a gray area between them is acknowledged in actual clinical practice. With these four distinctions, Yi Che-ma rearranges the *Treatise on Cold-Damage Disorders'* explanation of the six-channel pattern identification. For instance, to those who belong to lesser ŭm in Yi Che-ma's category, the *Treatise on Cold-Damage Disorders'* greater ŭm, lesser ŭm, reverting ŭm, greater yang, and yang brightness patterns of illnesses are relevant. In a similar vein, people of lesser yang in Yi Che-ma's framework are vulnerable to all the yang-related patterns in the *Treatise on Cold-Damage Disorders*. Among the greater ŭm people, to which group most Koreans belong, greater yang and yang brightness patterns of diseases from the *Treatise on Cold-Damage Disorders* are often found.[40] Although Yi Che-ma anticipated a completely novel conceptualization, his four categories of people reveal overlaps from the *Treatise on Cold-Damage Disorders'* six etiological identifications.

Not surprisingly, Yi Che-ma's classification of people reflects his accumulated observation of neighboring Koreans. He says, "Generally speaking, if we think of the proportion of 'greater and lesser' (*t'aeso*) and 'ŭmyang' people in a prefecture with a population of 10,000, greater ŭm people amount to 5,000, lesser yang to 3,000, lesser ŭm up to 2,000, and the number of greater yang is very small. Only between three or four and ten people belong to greater yang."[41] The four categories of people are not an imposed abstraction. Cases in *Longevity and Life Preservation in East-*

ern Medicine reflect Yi Che-ma's intention to deduce four categories from his clinical experiences.

Yi Che-ma also highlights how his own prescriptions are built on, yet depart from, past knowledge of herbs. For instance, he first picked up twenty-three formulas from the *Treatise on Cold-Damage Disorders*, which may well work for those who belong to the lesser ŭm people, then added thirteen formulas from the Song, Yuan, and Ming physicians. Finally, he suggested his own twenty-four prescriptions. For instance, Cinnamon Twig Decoction (*kyejit'ang*) was prescribed by Zhang Ji, Cinnamon Twig Plus Aconite Accessory Root Decoction (*kyeji pujat'ang*) was designed by Li Chan, and Yi Che-ma's solution was to build on the older formulas by adding milk vetch root or ginseng. Thus, he introduced Astragalus, Cinnamon Twig, and Aconite Accessory Root Decoction (*hwanggi kyeji pujat'ang*) and Ginseng, Cinnamon Twig, and Aconite Accessory Root Decoction (*insam kyeji pujat'ang*). In this way, he prepared more than one hundred of his own prescriptions.

An appreciation of Yi Che-ma's competence in his interpretation of the *Treatise on Cold-Damage Disorders*, particularly his deconstruction of the six-channel pattern identification, should be tempered with an acknowledgment of his dependence on Hŏ Chun. Yi Che-ma's organizing principle is partly consonant with Hŏ Chun's emphasis on the interaction between the surface sign and the interior root. Inasmuch as Hŏ Chun related the impact of cold damage to the internal landscape of the body, Yi Che-ma respected the evocation from the inner organs. For instance, his discussion about the lesser ŭm people is composed of three subsections, and their titles reflect the dynamics between dual relations. The lesser ŭm category is organized by the mode of heat or cold attack. Heat and cold reside in the exterior and interior, respectively, and are understood as combined with the corresponding internal organs. The other three categories of people are also conceptualized through the language of the exterior and interior, heat and cold, and outer and inner relationships, which are linked to corresponding organ manifestation.[42]

Not only the organizing principle and main terminologies but also textual components reflect Yi Che-ma's reliance on Hŏ Chun. Like Hŏ Chun, Yi Che-ma discloses his sources; the first few lines in the "discussion on the lesser ŭm people" begins with "Zhang Ji's *Treatise on Cold-Damage Disorders* said" or "Wei Yilin (1277–1347) said."[43] The remainder is filled with Yi Che-ma's identification of the works of Yuan and Ming China, such as Gong Xin's (active 1577–93) *Gujin yijian* (A mirror of medicine of all times, ca. 1589) or Zhu Zhenheng's *Danxi xinfa* (Danxi's methods of mental cultivation, 1481). Although he never directly mentions Hŏ Chun's *Precious Mirror of Eastern Medicine*, most of his Chinese originals were actually constructed by using Hŏ Chun's lines in the "cold" section of the *Precious Mirror of Eastern Medicine*. All of Yi Che-ma's Chinese references are already examined by Hŏ Chun. Yi Che-ma did not include any Chinese texts about the cold-damage disorders published after Hŏ Chun. Some of Yi Che-ma's sentences are much closer to Hŏ Chun's work than to the Chinese originals.[44] In other words, Yi Che-ma selected sentences from major Chinese literature using Hŏ Chun's text as a guideline.

YI CHE-MA'S IMAGINED LINEAGE AND AMBITIOUS SYNTHESIS

Given Yi Che-ma's departure from—yet reliance on—Hŏ Chun, how did he integrate his work with the notion of the tradition in which he was working? Yi Che-ma names the three most important figures in the entire history of medicine: Zhang Ji, Zhu Gong (1535–1615), and Hŏ Chun. Zhang Ji was viewed as having realized the way of medicine and left valuable writings. If Zhang Ji is the founding father of medicine, Zhu Gong restored medicine by expanding knowledge about disease patterns and prescriptions. After Song, Yi Che-ma viewed Yuan physicians such as Wang Haogu (ca. 1200–1264), Zhu Zhenheng, and Wei Yilin and Ming physicians such as Li Chan and Gong Xin as the legitimate successors of medicine.[45] The third most significant figure was Hŏ Chun. By juxtaposing Hŏ Chun with revered Chinese physicians, Yi Che-ma envisioned a lineage that is neither limited by geography nor hampered by political boundaries. By employing "Eastern medicine" in his title, he insinuated that the lineage first originated with Zhang Ji, was mediated by Hŏ Chun, and was to be succeeded by his own work. Except for this imaginative genealogy, no intact intellectual or cultural networks in his own locale are traceable.[46]

Yi Che-ma's ambition in compiling medical knowledge is also found in the scope of his text. Unlike his nineteenth-century contemporaries, who mainly worked on a variety of practical manuals, extracts of canonical texts, and popular guidebooks, he aimed at a grand synthesis that included the entirety of human existence. His four categories of people do not apply merely to human physiology. As Table 2 shows, the distinctions were to cover entire aspects of human life and unfolded differences in environmental frames, human affairs, virtues, wisdom, and emotional characteristics.

Table 2. Summary of Yi Che-ma's First Fifteen Sentences from
Sŏngmyŏnnon (Discussions on nature and order)

Environmental Frames	Times of Heaven (*ch'ŏnsi*)	Social Relationships (*sehoe*)	Human Relationships (*inmul*)	Locality (*chibang*)
(Sensory Organs)	Ears (*yi*)	Eyes (*mok*)	Nose (*pi*)	Mouth (*ku*)
Human Affairs	Appointed Roles (*samu*)	Social Acquaintances (*kyou*)	Clanships (*tangyŏ*)	Living Quarters (*kŏch'ŏ*)
(Internal Organs)	Lung (*p'ye*)	Spleen (*pi*)	Liver (*kan*)	Kidneys (*sin*)
Wisdom	Cleverness (*chuch'aek*)	Administration (*kyŏngnyun*)	Moderation (*haenggŏm*)	Generosity (*toryang*)
(Body)	Jaw (*ham*)	Chest (*ŏk*)	Navel (*che*)	Abdomen (*pok*)
Action	Discernment (*sikkyŏn*)	Nobility (*wiŭi*)	Craftiness (*chaegan*)	Scheme (*pangyak*)
(Body)	Head (*tu*)	Shoulders (*kyŏn*)	Waist (*yo*)	Hips (*tun*)

SOURCE: Yi Che-ma, *Sŏngmyŏnnon* 性命論, in *Tongŭi suse powŏn*. The table is from Ch'oe Sŭng-hun's translation.

Past scholarship has not shown an agreement in describing Yi Che-ma's intellectual background; rather, his fourfold conceptualization might be ascribed to his study of Mencius, to his exposure to Practical Learning, to his study of "warm disease" texts from Qing China, to Chosŏn elites' neo-Confucian pursuit of human nature, or simply to intellectual idiosyncrasy.[47] Given the manifold interpretations, what should be underscored is Yi Che-ma's desire to provide medicine with philosophical coherence. His detailed discussion about therapeutic principles comes after his teachings about "Nature and Order" (*Sŏngmyŏnnon*), the "Four Principles" (*Sadannon*), the "Establishment and Supplement" (*Hwakch'ungnon*), "Inner Organs" (*Changburon*), and the "Origin of Medicine" (*Ŭiwŏllon*). Furthermore, he left a series of philosophical writings prompting diverse interpretations and debates. To a certain degree, the grand intellectual scope and flexible mode of textual engagement made Yi Che-ma's *Longevity and Life Preservation in Eastern Medicine* attractive to later generations.[48]

Both Hŏ Chun and Yi Che-ma responded to the demands of their times. The medical world Hŏ Chun confronted seems to have shown only a superficial understanding of past knowledge and lacked sufficient depth to get to the root of the matter. King Sŏnjo, who supported Hŏ Chun's *Precious Mirror of Eastern Medicine*, complained that "recent medical texts both in China and Korea are all inconsistent selections from other copies, [and] thereby lack a consistent view."[49] Hŏ Chun attempted to create a synthetic order through which the full repertory of Chinese originals might shed light on Korean medical problems. Obviously, Hŏ Chun's contribution was to serve as the dynasty's paternalistic manager of medicine.

Conversely, Yi Che-ma learned and practiced medicine in a world where diversity and eclecticism were valorized more than ever. Medicine in general became more private and profitable during the nineteenth century.[50] The management of medicine gradually shifted from state regulation to private control. It was no longer surprising that a member of the *yangban* elite was acquainted with medical knowledge, managed seasonal herbs, and voiced his opinion when a local doctor was consulted for the illness of family members. Precious botanicals from Qing China were still in great demand, and the latest texts about materia medica, practical manuals for pregnancy and delivery, and specialized texts in pediatrics were popular. When Yi Che-ma wrote *Longevity and Life Preservation in Eastern Medicine*, a few Western-style clinics had already been established in Korea. Revisions of the *Treatise on Cold-Damage Disorders* that stressed empiricism over philosophical speculation prevailed, particularly in Japan starting in the eighteenth century. Paralleling these signs of change, Yi Che-ma clearly showed a divergence from the earlier Korean understanding of the treatise in terms of novel typology and self-awareness. However, his ambitious synthesis of the body, society, and morality was closer to the ideal of combining medicine with philosophy than to the new trend of dissociating them.[51]

REVISITING YI CHE-MA'S FOUR
CONSTITUTIONS MEDICINE

Surely, clinical efficacy was not the only reason that Hŏ Chun's and Yi Che-ma's syntheses of medicine became successful. During the eighteenth and nineteenth centuries, as was discussed earlier with the case of

New Compilation for Benefiting People, Hŏ Chun's *Precious Mirror of Eastern Medicine* was criticized and modified to meet novel demands, rather than blindly praised as an infallible native tradition. For more practical use, later generations ignored Hŏ Chun's teachings on "nurturing life," omitted some images, and abridged the original layout considerably. In a similar vein, Yi Che-ma's "Four Constitutions medicine" (Sasang ŭihak), taken from his four categories of people, was not complete from the beginning. During the early twentieth century, Yi Che-ma's framework was expanded and solidified as more practitioners came to add new herbs, formulas, and cases to the four categories.[52]

Wŏn Chi-sang's (1885–1962) *Tongŭi sasang sinp'yŏn* (Newly edited Four Constitutions of Eastern medicine, 1929) exemplifies the way Yi Che-ma's understanding of the *Treatise on Cold-Damage Disorders* was accepted by later generations of practitioners. Born to a *yangban* family, Wŏn Chi-sang highly regarded Yi Che-ma's interpretation of medical principles.[53] In the preface to Wŏn Chi-sang's work, Chang Pong-yŏng (1882–1948), Wŏn Chi-sang's friend, pointed out that Hŏ Chun's *Precious Mirror of Eastern Medicine* neither fully detailed the innate nature of internal organs nor elucidated the efficacy of useful herbs. Only Yi Che-ma, based on his categories of people, achieved an unprecedented degree of medical prowess by differentiating the characteristics of visceral organs and the technique of prescription.

Wŏn Chi-sang's text was primarily intended to reassemble Yi Che-ma's original. The work is composed of internal and external sections (*p'yŏn*), the internal sections having five subcategories and the external sections having four. The titles of all of the subcategories in the internal section include the term "four constitutions" (*sasang*). Even with this explicit aim to emulate Yi Che-ma's framework, Wŏn Chi-sang implicitly added his own modification. Primarily, he adopted the three layers of page division, which had already been used in Hwang To-yŏn's most popular medical primer, *Compendium of Prescriptions*, to effectively match prescriptions to relevant types of people. For each division by row, Wŏn Chi-sang put the three most prevalent categories of people. Above the three rows, he added another row to identify the names of illnesses. One illness name, in this structure, was related to three types of people vertically. Following the rows, readers were supposed to derive different prescriptions for each type of person.[54] In addition, Wŏn Chi-sang included "Prescriptions from Experience" (*kyŏnghŏmbang*) in the external section, aiming to insert his own clinical experience. His composition demonstrates that even though Yi Che-ma's four types of people were valorized, later practitioners did not hesitate to incorporate their own experiences of effective prescriptions and acknowledge the merit of a simpler layout for clinical efficacy.

The *Treatise on Cold-Damage Disorders* in Colonial Korea

NATIONALIZING EASTERN MEDICINE

The Korean understanding of cold-damage disorders during the early twentieth century began to depart from Hŏ Chun's and Yi Che-ma's line of terms and interpretations. Korean practitioners of traditional medicine began to combine these authors' accomplishments as sources of creating national and cultural identities of indigenous medicine while exploring diverse references from both China and Japan to compose the *Treatise on Cold-Damage Disorders* for a pedagogic purpose.

To further understand these changing contours, the colonial alteration of medicine, which primarily unfolded with the newly adopted licensing system, should be briefly explained. After the Japanese annexation of Korea in 1910, a series of regulations qualifying medical professionals were instituted in 1913.[55] The bottom line of the Japanese regulations of 1913 was to highlight the state's role in privileging Western medicine as the only authorized form of medicine, thereby disqualifying practitioners of traditional medicine. However, the colonial regime had more patience with traditional medicine in Korea than the government did in Japan. Labeled "apprentices of medicine" (*ŭisaeng*) by the Japanese, traditional Korean practitioners were tolerated but considered inferior to doctors of Western medicine, reflecting the government's intention to employ these "old-fashioned" agents of medicine in the Japanese colony.

Korean doctors of traditional medicine felt an urgent need to justify their careers under the government-driven degradation of the profession. Declaring Korea a unique place of medical development was a priority, thereby legitimating Korean doctors' infallible role in health management. Creating a nation-centered tradition of medicine enabled Koreans to uproot China as the source of medical authority. The nationalist framework endowed Korea with a main representative of the "East," which was enthusiastically contemplated as a way to relativize the "West." Leaders of traditional medicine who planned professional associations such as the Association for Studying Eastern and Western Medicine (Tongsŏ ŭihak yŏn'guhoe) and the Association of Traditional Medicine in Chosŏn (Chosŏn hanyak chohap) during the 1910s and 1920s wanted to legitimize traditional medicine with the nation-centered past. Instead of celebrating the origin of medicine with the Chinese Yellow Emperor and his servant Qibo, the continuity of medical tradition was represented by the legendary founding father of the Korean nation, Tan'gun, as well as Hŏ Chun's *Precious Mirror of Eastern Medicine* and the five hundred years of the Chosŏn dynasty. In the first volume of the monthly report of the Association for Studying Eastern and Western Medicine, a poem, in the form of a four-character couplet (*kasach'e*), celebrated the virtue of Eastern medicine and its long history.

> Hurray for our Eastern medicine
> It originated a long time ago and its root is deep
> From the Puso of the Tan'gun era
> The study of medicine began
> During the time of Kija, eight ancestors interacted
> Medicine became more serious
> After the import of medicine three thousand years passed
> The gradual development becomes evident
> Five hundred years passed after the beginning of the Chosŏn dynasty
> Studying medicine developed during the Chosŏn period
> One of the most significant achievements proved by history
> Is *Precious Mirror of Eastern Medicine*
> The virtue of Mr. Hŏ Chun
> Hasn't his beautiful name already been handed down to posterity?[56]

This Korea-centered imagining of medical development contrasts sharply with Yi Che-ma's recognition of and connection with medical innovators in Yuan and Ming China, not to mention Hwang To-yon's respect for the Yellow Emperor and Qibo as the origin of the universal medical principle.

Yi Che-ma's synthesis fits well into this demand for a Korea-centered articulation of medical heritage. His idiosyncrasy in combining medicine and philosophy provided a tool for Koreans not only to argue against the Western claim of scientific medicine but also to overcome the centuries-long authority of China. Cho Hŏn-yŏng (1900–88), a graduate of Waseda University in Japan (with a degree in English), became the most enthusiastic spokesman for traditional medicine during the 1930s. He took part in publishing a journal titled *Tongyang ŭiyak* (Eastern medicine) in 1935, and he introduced Yi Che-ma's work within a nationalist framework. In *Eastern Medicine*, Cho Hŏn-yŏng states,

> Yi Che-ma reminds me of Four Constitutions medicine and vice versa. Yi Che-ma is one of the greatest scholarly physicians of Chosŏn Korea. He is different from other Korean scholars who blindly admired what came from China. Yi Che-ma presented his own perspective on medicine, probed into an unexplored realm of medicine, [and] thereby argued for a novel synthesis of medicine. This is why we celebrate his achievement. Not only at a personal level, but for the entire field of Korean medicine, Yi Che-ma's accomplishment is a great pride. . . . It is a bit doubtful whether the infinite diversity of human affairs can squarely fit into those four categories. We may carefully examine this problem later. What should be underlined is that this theory was first claimed in Korea. Thus, every Korean practitioner of traditional medicine should understand this theory. This is why [our journal] introduces major points of Yi Che-ma's theory.[57]

Although Cho Hŏn-yŏng cast doubt on Yi Che-ma's rigid categorization, he never hesitated to celebrate his achievement, primarily because it was made by a Korean. The growing publicity around Yi Che-ma to a certain degree piggybacked on the nationalist will to invent a Korean tradition of medicine, which prevailed in the 1920s and 1930s. It was not until the middle of the 1920s that Korean advocates of traditional medicine in Seoul paid special attention to Yi Che-ma's work. Eventually, the Four Constitutions medicine gained popularity among Koreans starting in the mid-1930s.[58] When *Longevity and Life Preservation in Eastern Medicine* was recognized, the way Yi Che-ma connected himself to Zhang Ji and other Chinese scholarly physicians in an imaginative yet continuous line of transmission and revision of medicine became obsolete. Although Hŏ Chun's *Precious Mirror of Eastern Medicine* and Yi Che-ma's *Longevity and Life Preservation in Eastern Medicine* were accepted as canonical texts of Korean-style medicine, the linkage with the *Treatise on Cold-Damage Disorders*, which provides the core language and therapeutic references for Yi Che-ma's synthesis, became less acknowledged.

UNDERSTANDING OF COLD-DAMAGE
DISORDERS IN MODERN KOREA

While nationalizing Hŏ Chun's and Yi Che-ma's medical syntheses, Koreans continued to explore novel references to further understand cold-damage disorders. First and foremost, journals of traditional medicine that were published between the 1910s and 1930s served as a medium for spreading novel interpretations about the *Treatise on Cold-Damage Disorders*. Quite a few journals were intermittently published nationwide, and the extant volumes amount to more than fifty.[59] The earlier publications of the 1910s were often discontinued after one or two years due to financial instability, but circulation became more stabilized during the 1930s. Despite the different titles of the journals, their editorial boards showed continuity in terms of staffing and their networks. Most journals were published in Seoul, but *Ch'ungnam uiyak* (Medicine of Ch'ungnam), which was first published in 1935, aimed to represent local practitioners in the Ch'ungnam province. Because educational institutions of traditional medicine in Korea were rare and not standardized at the time, the journals primarily served as textbooks and newsletters, simultaneously reflecting Koreans' desire to revive and modernize their tradition of medicine. Articles ranged from passionate celebrations of the profession's long heritage to details about new medical regulations imposed by the Japanese colonial government, general information about Western medicine, a comparative glossary of traditional and Western medicine disease names, successful case studies, miscellaneous essays, and advertisements. In their instruction role, the journals included a series of lecture notes on cold-damage disorders in line with other subjects such as materia medica or gynecology.

The lecture series about cold-damage disorders became enriched as Koreans gained access to more diverse references. For instance, journals published between 1916 and 1919, such as *Tongsŏ ŭihakpo* (Newsletter of Eastern and Western medicine) and *Chosŏn ŭihakkye* (Association of Korean medicine), relied extensively on Tang Zonghai's (1862–1918) *Shanghanlun qianzhu buzheng* (Simple annotation and correction of the *Treatise on Cold-Damage Disorders*).[60] Tang Zonghai is widely known as the first proponent of medical eclecticism in the late Qing period, and his technique of synthesizing Chinese and Western medicine attracted Korean advocates. His other writings, such as *Zhongxi huitong yijing jingyi*

(The essential meaning of the medical canons [approached] through the convergence and assimilation of Chinese and Western [knowledge]) and *Jingui yaolue qianzhu buzheng* (Simple annotation and correction of the essentials from the *Golden Cabinet*), were presented to Korean audiences under the subject headings of "Treatise on Internal Organs" (*Changburon*) and "Studies on Miscellaneous Diseases" (*chappyŏnghak*). Koreans selected, copied, and summarized Tang Zonghai's interpretation of the *Treatise on Cold-Damage Disorders* in the midst of their growing anxiety about Western medicine.

Additional evidence of the eclectic interpretation of the *Treatise on Cold-Damage Disorders* is found in a text authored by Sŏng Chu-bong (1868–?) from the Ch'ungnam province. Titled *Hanbang ŭihak kangsŭpsŏ* (A textbook of traditional medicine, 1935), this book aimed to provide an effective guide for students who were supposed to complete their training of medicine within three years. Composed of 270 units (*kwa*) in six volumes (*kwŏn*), each unit required three days to master.[61] Under this planned curriculum, the author hoped the future generation would adapt to the rapidly changing world more smoothly. The author lamented that "recently, the fate of the world opened wide and the culture of the West and Asia developed respectively. Accordingly, hundreds of arts pursued the ultimate of refinedness. Alas, our profession of medicine did not fully come to grasp the urgency of the world, and merely wither[ed], discouraged, and did not fully reinvigorate."[62]

The cold-damage disorders as a teaching subject were central in Sŏng Chu-bong's composition of his textbook. The first volume outlines the general principles of ŭm and yang, the five phases, internal organs, and ki, relying mostly on Li Chan's *Introduction to Medicine*. The second and third volumes discuss the treatment of the cold-damage disorders. The fourth volume deals with miscellaneous diseases, and the fifth covers women's issues. Given his differentiation of the "cold damage" and the "miscellaneous disease" categories, Sŏng Chu-bong elaborated the discussion on cold damage by relying heavily on the original work of a Chinese scholarly physician, Huang Yuanyu (1705–58), particularly an explanation of warm disease.[63] Sŏng Chu-bong derived more than half of all his prescriptions from Huang Yuanyu's text, yet for the cold-damage section, almost every formula was taken from Huang's work.[64] Given Korean scholars' and practitioners' conventional pursuit of the latest medical texts from China, Sŏng Chu-bong's reliance on Huang Yuanyu is not surprising. What seems interesting here is that Huang Yuanyu's criticism of the overuse of attacking medicine in the case of warm diseases and other febrile epidemics parallels Korean practitioners' preference in prescribing harmonizing and replenishing prescriptions in treating cold-damage disorders.[65] Sŏng Chu-bong might have agreed with Huang Yuanyu in this therapeutic principle, then selected Huang Yuanyu's discussion about warm disease to enrich the "cold damage" sections of his textbook.

Koreans' interest in Chinese references continued into the 1930s, although Korean advocates enthusiastically aimed to displace China's authority and to establish their own origin and development of medicine. In particular, these advocates were attracted by China's ostensible attempts to modernize traditional medicine. One of the leading Korean voices, Sin Kil-gu (1894–1972), eagerly introduced the details of new educational institutions, journals, and various publications of traditional medicine in China. For instance, he reported that Yan Xishan (1883–1960), a prominent warlord in Shanxi, came to support traditional medicine after he successfully recovered from a skin disease with the help of a traditional physician. The traditional medicinal community in China, according to Sin Kil-gu, had more than seventeen educational institutions in major cities, such as Shanghai, Suzhou, and Kaifeng, and more than thirty periodicals specializing in traditional medicine. Doctors of traditional medicine in China seemed to be vigorously competing with Western-trained doctors without losing popularity and self-confidence. To Korean doctors of traditional medicine, their Chinese counterparts appeared to be more autonomous because they were not hemmed in by the strict licensing regulations of a colonial government. It was also enviously reported that Chinese scholars had proudly edited the voluminous work *Zhongguo yixue da zidian* (The great dictionary of Chinese medicine) and had even begun to historicize their tradition by publishing *Zhongguo yixue shi* (The history of Chinese medicine), written by Chen Bangxian (1889–1976). In other words, Korean doctors of traditional medicine remained aware of their marginality, not only with respect to the West but also within East Asia with respect to China.[66]

In the 1930s, the Korean understanding of the *Treatise on Cold-Damage Disorders* reflects the growing hegemony of Japanese academia in defining the novelty and authority of medical knowledge. Under the Japanese-led modernization of medicine, Korean doctors of traditional medicine could not avoid relying primarily on Japan as a point of reference. For instance, Korean advocates of traditional medicine regarded the establishment of the Japanese Society of Traditional Medicine (Nihon kanpō igakkai) as a positive sign of a revival of traditional medicine. The fact that Western medicine–trained Japanese scholars took part in this society encouraged Koreans to argue for the significance of traditional medicine as an intellectually appropriate therapeutic principle. The society's journal, *Kanpō to kanyaku* (Formulas and herbs of traditional medicine), first published in 1934, was often quoted by Koreans as evidence of the scientifically proven efficacy of traditional medicine.

The Korean study of the *Treatise on Cold-Damage Disorders* followed this trend. In the journals published and circulated in the 1930s, articles on cold-damage disorders abound, particularly those written by Japanese authors, such as Yumoto Kyūshin (1876–1941) and Yakazu Dōmei (1905–2001).[67] In particular, Yumoto Kyūshin was recognized for his work in adopting and revising Yoshimasu Tōdō's (1702–73) radical reinterpretation of the *Treatise on Cold-Damage Disorders*. Rejecting the neo-Confucian interpretation of medicine, Yoshimasu Tōdō had argued that all diseases could be ascribed to one kind of poison. Hence, treatment should be aimed at attacking the noxious pathogen that has accumulated in the body, which can be detected by a refined diagnostic palpation of the abdomen.[68] Inheriting but partly overcoming this radical empiricism, Yumoto Kyūshin aimed to revise the Ancient Formula Current's (Kohōha) approach to the *Treatise on Cold-Damage Disorders*, and his synthesis, *Kokan igaku* (Sino-Japanese medicine, 1934), was circulated in China and Korea and then translated into Korean in the 1960s.

Of course, some Korean reformers of traditional medicine criticized the Japanese emphasis on the Ancient Formula Current's understanding of the *Treatise on Cold-Damage Disorders*. Sin Kil-gu states,

> They act as if there were no other theories except Zhang Zhongjing's *Treatise on Cold-Damage Disorders*. Thus, they indiscriminately apply the principle of cold damage, which is supposed to be applied only to disorders of external origin, to miscellaneous diseases caused by inner factors, [and] thus devote all their energies in attacking and purging. As such, they rashly prescribe *Ephedra sinica* [*mahwang*] and *Pinellia ternata* [*panha*] for tuberculosis.[69]

Korean advocates ascribed Japanese scholars' inclination toward strong remedies to the general mind-set of the time. Inasmuch as Westernized medicine preferred the attacking principle to more gentle approaches, traditional medicine also came to be influenced by this trend. However, according to Sin Kil-gu, this is nothing but a regrettable example that reveals the Western impact on the traditional. He expected this tendency to be corrected before long. Despite this criticism, it is not an exaggeration to say that Japanese writings in general were circulated among Korean doctors of traditional medicine as references to further understand the *Treatise on Cold-Damage Disorders.*[70]

Conclusion

The Korean understanding of cold-damage disorders over time partly mirrors the textual component of Eastern medicine. Hŏ Chun's passages about cold damage make up the bulk of his description of miscellaneous diseases and shrewdly display his technique of balanced composition, which became a crucial ground for conjuring up Eastern medicine. Hŏ Chun's composition of cold damage continued to provide textual resources for Yi Che-ma in the late nineteenth century, allowing Yi Che-ma to substantialize his four categories of people; this was developed as Four Constitutions medicine under the nationalist reframing of indigenous medicine in the early twentieth century. In colonial Korea, advocates of traditional medicine frequently used Eastern medicine as a label to mark the territory of its own history and as an institutional initiative for decentering Chinese authority. However, in fleshing out modern educational materials on traditional medicine, Koreans needed more engagement with, not separation from, Chinese and Japanese references. While proposing the *Treatise on Cold-Damage Disorders* as an indispensable subject of education, Koreans relied on a Chinese proponent of medical eclecticism, Tang Zonghai; Huang Yuanyu's study of warm disease; and the work of Yumoto Kyūshin, a Japanese expert on the *Treatise on Cold-Damage Disorders.*

Taking into account the Korean use of the *Treatise on Cold-Damage Disorders*, the imagined geographic distinction of the East solidified no parochial inclination. Although Hŏ Chun and Yi Che-ma highlighted Eastern medicine for an intellectual and clinical initiative, neither valorized regionalism in medicine by documenting specific disease patterns or a set of indigenous therapeutic resolutions. In addition, Eastern medicine in the court-sponsored anthology in sixteenth-century Seoul aligns partially with Yi Che-ma's idiosyncratic synthesis in the marginalized northeast of nineteenth-century Korea. Their connection, through the label of Eastern medicine, was solidified with the nationalist turn in the early twentieth century. The use of Eastern medicine demonstrates each author's will to name his own intellectual cornerstones, mapping the contours of the surrounding medico-intellectual world on his own terms. In Korean textual composition, "the East" signifies not so much a region in reality as an intellectual standpoint in motion. Conjured up against Southern, Northern, and Western counterparts, Eastern medicine enabled Korean elites to name their positionality for medical innovation. The deliberate demarcation of Eastern medicine reveals more hybridity than purity in its textual, clinical, and cultural components. Although the growing nationalization of medicine in the twentieth century highlighted Eastern medicine's own territory in contrast to its Chinese, Western, and Japanese counterparts, Korean physicians have simultaneously sought connections by engaging different sources across national, cultural, and linguistic boundaries to effectively meet the clinical and social demands of their time.

CHAPTER THREE

Chosŏn Koreans

The Colonial Identification of the Local

On December 31, 1934, Kim Yŏng-hun (1882–1974) published an article in *Chosŏn ilbo* (Korean daily) arguing for a Chosŏn (Korea)-centered medicine.[1] He firmly believed health and diseases were intimately associated with indigenous ways of life, and thus the principles of diagnosis and prescription should be defined by Korean traditions of eating, clothing, and bedding. Criticizing the recent domination of "Western medicine" (*sŏŭihak*) in policy and education, Kim Yŏng-hun urgently called for a medicine firmly rooted in "Chosŏn Korea's own cultural and socioeconomic situation." Challenging Western medicine's claims of superiority, he advocated for medicine that is "most convenient for curing Chosŏn Koreans, most accessible, and most ideal for the specific conditions of the East (*tongyang*)."

Kim Yŏng-hun's call for a Chosŏn-centered medicine resonated with the emerging voices of a reviving traditional medicine in the 1930s. A group of advocates debated with doctors of biomedicine in a series of articles presenting pros and cons of each side, which were published by the *Korean Daily* and other print media from 1934 to 1937. Kim Yŏng-hun's vision of a new medicine shared the ideals and strategies of a group of advocates seeking to legitimate indigenous medicine.

The controversy to which Kim Yŏng-hun was responding arose primarily from questions regarding the universal validity of Western medicine. Should the principles of medicine be uniform, transcending any differences of place and human constitution? Or is the principle of medicine firmly rooted in the local situation, and if so, should plurality be assumed as intrinsic to the learning of medicine? Answering these questions involves tackling issues around translation and hierarchy. If one universal principle governs, then the local differences in conceptualizing health and sickness should be succinctly translated into the universal grammar of medical science. On the opposite end, if a pluralistic medical standard is assumed, multiple representations of health and diseases would be permitted and the ordering of the different modalities of medical heritages should thereby be negotiated. By considering Kim Yŏng-hun's Chosŏn-centered medicine and related controversy in the 1930s, this chapter traces different solutions to those raised questions. In what ways did the particular nature of Chosŏn Korea become a significant object of medical discourse among different professions? What role did the specificity of Korea play in conjuring up intellectual and institutional grounds for a new medicine?

In addition to Korean advocates of traditional medicine, Japanese, as teachers and practitioners of biomedicine, produced substantial research about the uniquely Korean nature of health and diseases. During the controversy, Korean advocates of traditional medicine selectively relied on the latest Japanese medical research about Chosŏn Korea. For instance, Japanese scholars examined the medical efficacy of Korean herbs, moxibustion, and acupuncture. A series of articles published in *Chōsen igakkai zasshi* (Journal of the Chōsen Medical Association) reveals the increasing interest of Japanese researchers in biomedically examining the therapeutic potential of traditional medicine.[2] Published by Japanese doctors who were affiliated with the Government-General Hospital in Korea (Chosŏn ch'ongdokpu ŭiwŏn), Japanese-run provincial hospitals, and the Keijō Imperial University (Keijō teikoku daigaku), the journal displays the convention of "scientifically" reporting on the unique traits of the bodies of Koreans and the specific conditions of health and illnesses in Korea.[3]

Paralleling the journal's overall growth, the Japanese researchers increasingly articulated the category of "Chosŏn." Beginning in the 1920s, articles that included "Chosŏn" in the title or used it as a unit of analysis increased, reaching their peak in the late 1930s. From the chemical analysis of ginseng to the survey of parasites prevalent among Koreans to the study of the physique of Korean prisoners, biomedically trained Japanese explored the embodied nature of "Koreanness." Partly in line with the growth of racial science in the 1930s, Japanese researchers augmented their reports about Chosŏn Korea. Taking the *Journal of the Chōsen Medical Association* and related journals as examples, this chapter dissects the terms, grammar, and rhetoric of the Japanese biomedical portrait of Korean distinctiveness and its dissemination and circulation among a Korean readership.

The significance of the Japanese medical discourse lies in how it shaped the framework of scientific research carried out by Korean doctors of biomedicine. From the 1920s on, newly emerging Korean professionals of biomedicine joined in constructing the scientific depiction of the Korean body. For instance, the *Journal of the Chōsen Medical Association* includes fewer than five articles by Korean authors between 1911 and 1926 but more than two hundred by Korean authors between 1926 and 1942. Not all of them use the ethnic category for analysis, yet a substantial number delve into the idea of Korean distinctiveness in medicine. Since most eminent Korean doctors of biomedicine studied with Japanese teachers, Koreans emulated the Japanese research framework, which had itself largely replicated the German model.[4] Biomedical terms, statistical methods, and Western styles of academic writing were used to strengthen Korean doctors' authority in representing Korea's specificities in Japanese. With the same research objects and methodologies, Korean depictions of the self revealed differences from their Japanese counterparts in rendering data scales and carrying out deep analysis.

Analyzing the three groups' growing enthusiasm for articulating the Chosŏn label, this chapter considers those multiple agents' divergence and connectivity in authenticating indigenous attributes in the 1930s. Obviously the newly imposed license system defined clear boundaries, creating a sense of territory surrounding the realm of authority assigned to doctors of biomedicine and traditional medicine. Yet more significant, the medical knowledge produced by each realm intersected, creating unexpected room for emulation and contestation. This chapter analyzes the evolution of the 1930s' medical discourses about the Chosŏn Korean traits through which medical elites fashioned their professional, cultural, and ethnic identities.

By examining the three groups, this chapter raises the "new questions" that Projit B. Mukharji presents after reviewing recent scholarship about Western medicine in a colonial context in Asia. Whereas old questions sought to specify "clear-cut domination and subordination," new inquiries have fleshed out the formative moments and mechanisms through which seemingly stable categories such as "native," "Western," "indigenous," and "biomedicine" were constantly reestablished and modified. While analyzing the process of the Indian accommodation of

Western medicine, Mukharji attempts to excavate the ways "'western' medicine 'produce(s)' new kinds of subjects, identities, spaces, objects, etc., and how these came to be organized within a complex and shifting matrix of Power and Resistance."[5] The circulation of "Chosŏn" as a unit of analysis in the 1930s exposes the underlying viewpoints, motivations, and conduits of relationship that shaped the meaning of the indigenous. By examining the socially situated traffic of medical discourse, this chapter analyzes the possibilities and limits of biomedicine in producing knowledge of the ethnic specificities of Chosŏn Korea in a colonial setting. Accelerated circulation of the Chosŏn label in the 1930s not only solidified the "Korean" distinction but testified to its instability as well.

Korean Practitioners of Traditional Medicine

The controversy in the 1930s exposes the extent to which Kim Yŏng-hun's vision of Chosŏn-centered new medicine was intertwined with the manifold challenges of biomedicine.[6] The debate was triggered by an article written by Chang Ki-mu (n.d.) in February 1934 arguing for the revival of traditional medicine.[7] Chang Ki-mu's call can first be read as a practical proposal for modernizing traditional medicine: he wanted to create professional associations, research centers, public lectures, and media support to improve the way traditional medicine was practiced in Korea. A biomedicine-trained doctor, Chŏng Kŭn-yang (ca. 1912–?), strongly opposed rejuvenating traditional medicine.[8] As each side developed its own claim, Cho Hŏn-yŏng (1900–88), Yi Ŭr-ho (1910–98), Sin Kil-gu, and Kang P'il-mo (n.d.), whose careers were related to either traditional medicine or biomedicine, contributed to the debate. In particular, Cho Hŏn-yŏng struck a sensitive chord while persuasively and enthusiastically arguing for validating, protecting, and institutionalizing traditional medicine in Korea. Until 1940, Cho Hŏn-yŏng repeatedly promoted the improvement of the overall status of traditional medicine for Chosŏn Koreans.

PROVINCIALIZING BIOMEDICINE

One of the most noticeable issues dividing the two parties was how to reconcile universalist and pluralist perspectives. Unlike the advocates of traditional medicine, who assumed a fundamental difference between traditional medicine and biomedicine, Chŏng Kŭn-yang viewed the claim that systems of medicine are plural as sectarian and outmoded. According to him, only ignorance of science nurtures sectarian partiality. Based on his belief in universal science, he refused to validate plural systems of knowledge, in which cultural heritages or regional distinctions might legitimate the significance of the particular standpoint. In Chŏng Kŭn-yang's view, the development of medicine aimed for universal progress, and what Korea needed was to move toward the scientific universal, not get dragged down into the inferior particularism. He evaluated the efforts of traditional physicians as nothing but an attempt to promote their private and selfish interests.[9]

Chŏng Kŭn-yang's opponents, however, rejected biomedical universalism. For instance, Chang Ki-mu supported pluralism in medical understanding, noting that if one ignores (Chosŏn-centered) traditional medicine only because it does not fit into biomedicine, this will eventually hamper the ultimate progress of medicine.[10] The universal medicine Chang Ki-mu conjured up contains both Western and Eastern components and other unknown schools of healing. Localizing biomedicine as an influential yet limited component of the universal, Chang Ki-mu described the success of Western medicine more as a contingent achievement than as an inevitable step toward universal progress. In general, he thought that medicine developed either by logical inference or by good luck. What happened in Western medicine was the contingent combination of these two factors.[11]

Building up Chang Ki-mu's argument, Cho Hŏn-yŏng tried to identify biomedicine as the opposite but complementary partner of traditional medicine. Like the balanced polarization of ŭm and yang as a whole, the two medical systems were envisioned as interdependent, each one an indispensable partner in perfecting medicine to some degree. One could not replace the other. The polarized differences may have produced tension, but so did the possibility of collaboration. Cho Hŏn-yŏng elaborated that Eastern (Chosŏn-centered) and Western medicines served as "synthetic versus sectional remedies," "natural versus artificial healings," "dynamic versus static medicines," "healing focused on the original cause versus presenting symptoms," "supportive versus aggressive medicines," "internal versus external treatments," "adaptable versus rigid medicines," "medicine of commoners versus that of aristocracy," and "official versus private medicines."[12]

Given the polarization, traditional medicine was indispensable in redressing the deficiencies of Western medicine. By pointing out the temporality and particularity of Western medicine, the supporters of traditional medicine projected a novel ground for a universal medicine that stemmed neither entirely from Western novelty nor from Eastern heritage. Under the sociocultural pressure of being rendered obsolete, Cho Hŏn-yŏng and other advocates envisioned an ideal medicine in which the laws of Western natural science are just one component of treating health and diseases.[13]

As a way of relativizing biomedicine, supporters also emphasized the historical mutability of disease, and thus the variability of medical responses. The constantly evolving nature of diseases was supposed to authorize only temporal and local intervention. According to Sin Kil-gu, "Traditional medicine from the ancient era of *Divine Husbandman* and the *Yellow Emperor* down to the present has been continuous for five thousand years. Many wise people and scholars took account of changes in local characteristics, class, heredity, human nature, climate, nature of soil, customs, and life style."[14] Overall, the atmosphere of a specific era determined the characteristics of a medical trend, thereby explaining the changing tones and attributes of medicine over history. According to the "civil and military" or the "royal and military governing" of the times, clinical intervention also aligned with the principle of "replenishing yang" (*poyang*) or "attacking and purging" (*kongsa*). The differences between those two seemingly contrary principles were interpreted as the manifestation of dialectic.[15]

Given the close relationship between a specific attribute of an era and its healing principle, Cho Hŏn-yŏng comprehended the dialectic changes among diverse currents in medicine. Reflecting the tenor of the times, such as harsh and martial or mild and civil, famous Chinese physicians Liu Hejian (ca. 1186) and Zhu Zhenheng were classified as patrons of "cold and cooling medicine"; Li Gao, Li Chan, and Hŏ Chun as "compromisers"; and Zhang Jiebin (ca. 1563–1640) as "warming and replenishing." This way of conceptualizing the past redefines the changes within tradition and offers a pattern for understanding the present domination of Western medicine. According to the correlation between medicine and the times, the 1930s could be understood as a military age and a time for purging therapies. The advance of Western medicine in line with Western civilization augmented competition and confusion, resulting in increased symptoms of irritation, anxiety, and nervousness.[16] Without knowing how symptoms were related to the attributes of the times, many members of the Korean public were blindly seeking fast solutions from harsher and stronger remedies, where Western medicine showcased its prowess.[17] By categorizing the entire history of medicine according to the alternating influence of contrary principles, advocates of traditional medicine gained a framework through which the dominance of biomedicine could be historically relativized.

The ideal balance between East and West, or the domestication of Western medicine, had to accommodate the Chosŏn political and cultural reality. Even the most passionate advocate of traditional medicine believed it could hardly survive without translating its major principles into the terms and reasoning of Western medicine. Cho Hŏn-yŏng and other colleagues felt that they lived in a world where the "Western influence occupied the East" (*sŏse tongjŏm*) politically and intellectually. Given the biased power framework, supporters of traditional medicine actively borrowed examples and analogies from its counterpart to legitimate the medical efficacy of their profession.[18]

One good example of this tendency is Cho Hŏn-yŏng. He competently employed biomedical examples to demonstrate that experts in traditional medicine were far from being ignorant and old-fashioned.[19] Based on his knowledge of anatomy and physiology, Cho Hŏn-yŏng explained why diabetic symptoms could be better treated by traditional medicine. Brown seaweed, well known as an aid to the recovery of women after childbirth, was efficacious due to the function of iodine. Acupuncture and moxibustion, often despised by doctors of Western medicine, were compared with chiropractic or osteopathy as working on the nervous system. Similarly, acupressure therapy aided blood circulation and electrochemical transformation. Moxibustion and heat-acupuncture therapy were described as effective because of the interaction of white and red blood cells and digestive enzymes. Cho Hŏn-yŏng also argued that the development of endocrinology would eventually help the doctors of Western medicine fully appreciate the existence of *ki* and its circulation system (*kyŏngnak*).[20] Cho Hŏn-yŏng was a popularizer who was able to evoke the scientific to give a modern tenor to traditional medicine.

In the debate, most advocates of traditional medicine were well-educated young intellectuals. Compared to the advocates of traditional medicine in the 1910s, supporters in the 1930s were certainly empowered by their familiarity with biomedicine. For example, Chang Ki-mu was one of the graduates of the Government School of Western Medicine (Kwallip ŭihakkyo) founded in 1899. Although he intended to be a doctor of biomedicine, he was drawn to traditional medicine as he realized the limits of the "scientific" approach to health and illnesses. Yi Ŭr-ho was a graduate of the Professional School of Pharmacy at Seoul (Kyŏngsŏng yakhak chŏnmun hakkyo), which aimed to produce biomedicine-trained pharmacists, but he was also interested in traditional medicine and became an apprentice to an experienced master of it. Finally, Cho Hŏn-yŏng majored in English at Waseda University in Japan and was known to be actively involved in the Association of Korean Students in Japan.[21]

For these newly educated supporters, Western medicine never fully demonstrated its superiority to the millennium-old experience of traditional medicine. The advocates' modern education enabled them to extend criticism of biomedicine more persuasively. These men thought the efficacy of Western medicine was proven only in the finesse of surgical treatment, better-controlled epidemics, and enhanced hygienic regulations. However, traditional medicine was superior for chronic diseases, internal medicine, and as an alternative to surgery. Using their knowledge of biomedicine, these defenders continued to characterize Western medicine as reductionist, artificial, mechanical, and lacking the holistic and natural perspective indispensable to understanding the living organism. According to them, fast treatment, which was known as the strength of Western medicine, could not be the highest standard for appropriate cures; rather, the underlying concepts of diagnosis and the logistics of treatment should be fully evaluated; in this sense, traditional medicine was hardly inferior to its counterpart. Yi Ŭr-ho stated,

Most followers of the mechanistic view of medicine belong to the school of Pasteur, and tend to define health as a germ-free state, which is too narrow to fully grasp the true meaning of health. Although disease is not entirely irrelevant to a bacterium, if a disease is exclusively ascribed to a certain bacillus, then exaggerated fear about bacterium will be predominant. More to the point, this emphasis on germs might establish medical therapy only by attacking the germs, which eventually degrades the natural immunity of a human body. A bacterium is nothing but a kind of diagnostic symptom that is specialized to a certain disease. The Western medicine's definition of health and disease was centered on the outside condition that includes a certain type of bacterium, thereby mostly resulted in the removal of a locally specific lesion or overuse of germicides. This not only ignores the natural attributes of a human body but also destabilizes the balance of physiological function. To achieve a more perfect therapy, isn't it necessary to avoid a hasty removal and overuse of bactericide and to respect the physiological autonomy of a body?[22]

This reasoning, which may seem partially plausible to contemporary audiences in terms of the extended definition of health, represented an urgent defense of traditional medicine in the face of biomedicine's belief in the ultimate authority of experimental evidence as the only authentic method of understanding bodily functions.[23] Here the traditional medicine supporter not only used biomedicine to demonstrate a modern outlook but also presented himself as a knowledgeable critic of the philosophy of biomedicine. Although it is difficult to measure which side prevailed in the debate, the advocates of traditional medicine evidently struck a chord among the educated Korean audience.[24]

These advocates' enthusiastic efforts to craft a dialogue with Western medicine eventually gained empathetic audiences. A biomedically trained doctor, Kang P'il-mo, published an article in *Tonga ilbo* (East Asian daily), explaining,

> According to my last four-year experience, chronic infirmities, such as pulmonary tuberculosis, pleurisy, neurasthenia, chronic dyspepsia, which have no specific therapy, are treated better and faster by traditional medicine. Acute diseases, such as influenza, acute tuberculosis, acute pleurisy, pertussis [whooping cough], measles, typhoid fever, and dysentery once seemed very serious. However, I have experienced that those acute diseases are often well treated with a couple of packages of prepared herbs. Although traditional medicine doctors regard these cases as normal, yet to a doctor of biomedicine like me, the great truth of traditional medicine seems just amazing. And this makes me realize the limits of modern medicine [*hyŏndae ŭihak*], particularly the constraint of internal medicine.[25]

Kang P'il-mo's positive evaluation of traditional medicine might not have represented the ideas of the majority of his biomedical peers. However, other Korean doctors of biomedicine in the 1930s were reported to have begun scrutinizing the medical efficacy of traditional medicine. Even Chŏng Kŭn-yang, the determined representative of Western medicine in the debate, participated in a public lecture aimed at discussing the biomedical effectiveness of traditional medicine. Sin Kil-gu, another advocate of traditional medicine, commented later that Chŏng Kŭn-yang's attitude became more positive and balanced compared with the antagonism he had expressed during the 1934 debate. Sin Kil-gu took Chŏng Kŭn-yang's participation in the public lecture titled "The Path toward Synthetic

Medicine" as evidence of the popularity of traditional medicine among biomedical experts.[26]

THE ASSESSMENT OF THE
MEDICINE MARKET

In addition to the aforementioned agendas, the debate showcased a conflict over regulating physicians' fees and the price of medicinal drugs, which involved ethical issues as well. To doctors of traditional medicine, Western medicine only served the rich. Due to the high cost of education and extravagant facilities, Western medicine could not be accessible to ordinary Koreans. On the contrary, traditional medicine was viewed as being more egalitarian and popular, which allowed for a variety of approaches.[27] Doctors of biomedicine considered the low cost of traditional medicine not as an intrinsic strength but as a temporary advantage that the comparatively low social status of traditional medicine created at that juncture.[28]

Sales of medicinal drugs (*maeyak*) added more controversial issues. Whereas doctors of biomedicine said that traditional medicine had a muddled drug market, doctors of traditional medicine blamed their Western counterparts for accelerating commercialization by standardizing medicine. Both claims captured the eclectic and perplexed circumstances of the drug market in early twentieth-century Korea. Chŏng Kŭn-yang, a doctor of biomedicine, regretted that medicinal drugs were consumed by lay people without any qualified guide or medical advice. People self-diagnosed and self-prescribed without consulting medical experts. To him, marketing medicine was purely for profit and was thus not even comparable to the practice of "real" medicine. By likening the populist feature of traditional medicine to an unskilled and crude pursuit of profit, Chŏng Kŭn-yang trivialized his counterparts' viewpoint.[29]

In response to this criticism, patrons of traditional medicine criticized the overall trend of standardization in biomedicine. It was hard to commercialize traditional medicine due to the diverse names of herbs and individualized prescriptions. Unlike biomedicine, each patient's symptoms and economic situations were supposed to be reflected in the personalized prescriptions of traditional medicine. Biomedicine simply matched diseases with prescriptions, ignoring the subjective experience of illness. The prescription of standardized medicine should be blamed for accelerating the commercialization of medicine.

Finally, Cho Hŏn-yŏng expressed regret at the disappearance of literati doctors (*yuŭi*).[30] As a result of rigid professionalization, traditions of scholarly medicine gradually faded away. Cho Hŏn-yŏng suggested that the heritage of literati doctors should be renewed to establish a more democratic and populist approach to medicine. Those who wanted to become literati doctors should be tested for intellectual ability and moral character. Unlike the rigid qualification of medicine the colonial government had created, Cho Hŏn-yŏng envisioned more layers of practitioners, where boundaries between professions would be flexible. More than anything else, he opposed medical monopolies by means of rigid license regulation. Addressing examples from contemporary Tokyo and Osaka, where hundreds of nonlicensed practitioners practiced medicine without causing any serious problems, Cho Hŏn-yŏng argued for liberalizing traditional medicine and freeing it from hierarchical and monolithic institutionalization.

The debates in the 1930s did not induce any radical changes in colonial health policy.[31] They did expose a vision for indigenizing medicine that recognized the benefits of scientific medicine from the West but simultaneously aimed to go beyond its superiority claim institutionally and epistemologically. It is no surprise that Cho Hŏn-yŏng moved to North Korea after 1948, seeking a new path to realize the nationalist, egalitarian, and populist management of health in the newly emerged communist regime. In the south, too, the vision of Chosŏn-centered medicine that surfaced during the 1930s was revisited by Korean practitioners of traditional medicine in the 1990s when they faced another crisis caused by biomedicine's expanded territorial claim.[32]

The Korean yearning for indigenous initiative established the polarized sphere of Chosŏn's own medicine as an antithesis of biomedicine. To flesh out the content of Chosŏn-centered medicine, advocates actively used biomedical terminologies and theories; emphasized the temporality of diseases and clinical principles; and conjured up a nonprofessional practice of medicine outside the monopoly of state and market. In this process, advocates of traditional medicine selectively referred to the latest Japanese research about Chosŏn Korea and traditional medicine. The next subsection thus examines the way Japanese medical research used the label of "Chosŏn," thereby inspiring a Korean readership.

Japanese Views of Korean Bodies

BIOMEDICINE BY JAPANESE AGENTS
IN KOREA

Japanese medical research about Chosŏn Korea reached its peak in the 1930s. While cataloging names of herbs, investigating patterns of diseases, and quantifying Korean bodies, Japanese medical professionals elaborated the idea of Koreanness. Focusing on two major medical journals and related Japanese writings, I scrutinize the evolution of Japanese research conventions, through which the newly emerging Korean biomedical doctors also learned how to scientifically portray the indigenous.

The overlaps and dissimilitude between Japanese and Korean reports of the "Chosŏn" help us redress the conventional ideas about colonial medicine in Korea. Across disciplines and genres, Japanese management of biomedicine in Korea has been judged negatively. For instance, the memoir of Kim San (1902–34), a well-known Korean communist who joined the Chinese Revolution, noted the oppressive practice of vaccination by Japanese police.[33] Rumors said that Yun Tong-ju (1917–45), after his death the most lauded poet in South Korea, perished due to the Japanese injection.[34] Korean historians highlighted the limitations of the Japanese-run biomedicine: the newly opened Japanese private clinics of Western medicine at treaty ports mainly received Japanese patients; provincial hospitals run by the government-general aimed to enhance colonial control, ignoring the overall innovation of public health in Korea; and the sanitary police harshly imposed hygienic regulations on the bodies of colonized Koreans.[35] More to the point, in 1916, a Japanese professor of anatomy, Kubo Takeshi (1879–1921) discussed the scientifically proven inferiority of Korean bodies, which caused Korean students to protest.[36]

To a certain degree, the oppressive nature of biomedicine, mediated through Japanese agents, coincides with what Ruth Rogaski describes as the "dual nature" of biomedicine in a colonial setting. Analyzing the Japan-led exercise of hygiene (*weisheng*) in Manchuria during the first half of the twentieth century, Rogaski highlights the fact that the longing to be modern through the practice of biomedicine and hygiene was unavoidably accompanied by coercion. She underscores that "health and violence, cleanliness and coercion, religious benevolence and scientific objectification" coexist and sustain the modern premises of public health and biomedicine.[37]

The double-edged attributes of colonial medicine encourage us to delve into the interaction that biomedicine facilitated between Japanese and Koreans. The Japanese portrayal of the indigenous did not merely objectify the Korean Other but also created room for a group of Koreans to formulate their longing to become scientific and professional. Two medical journals in Japanese are scrutinized in the following analysis: the *Journal of the Chōsen Medical Association*, which was published by the Society of Medicine in Korea (Chōsen igakkai), and *Mansen no ikai* (Medical world of Manchuria and Korea), which was supported by a well-known Japanese apothecary. The Society of Medicine in Korea was organized by Japanese physicians who were mostly affiliated with the Department of Medicine at the Keijō Imperial University. Founded in 1924, the Keijō Imperial University was the most prestigious academic institution outside Japan for Korean medical students. The Medical Department not only transmitted the latest Western medical theories from the Tokyo metropolis to the colony but also translated what was particular and indigenous in Korea into a universal grammar of science. *Medical World of Manchuria and Korea* had a layout similar to that of the *Journal of the Chōsen Medical Association*, including research articles, news, and advertisements.[38]

Although circulation is difficult to measure, each journal served as an authorizing medium and communicative sphere for professionals from each regional organization of biomedicine. For instance, professional meetings of the College of Medicine in Manchuria (Manshū ika daigaku) and public lectures and local meetings held in major cities of Manchuria, such as Fengtian or Fushun, were frequently reported. In addition, the entrance examination test questions and the list of successful candidates of the Professional School of Medicine in Keijō (Kyŏngsŏng ŭihak chŏnmun hakkyo) were introduced.[39] Medical activities in major cities of Korea, such as P'yŏngyang, Taegu, and Pusan, were reported, as well as news from the Society of Otorhinolaryngology in Japan and the Society of Gynecology in Japan.[40] All in all, *Medical World of Manchuria and Korea* and the *Journal of the Chōsen Medical Association* helped professionals of Western medicine in Korea relate their knowledge and activities to an experts' community that transcended regional, cultural, and ethnic boundaries.[41]

Most authors who published in the journals were Japanese natives in the 1920s, whereas coauthorship between Japanese and Koreans began to emerge during the late 1920s, and then became frequent in the 1930s. Korean authorship also rose during the 1930s. During the 1930s, the two journals often published six or eight research articles in each volume, among them two or three articles specified as scientific investigations of "Korean" traits. *Medical World of Manchuria and Korea* and the *Journal of the Chōsen Medical Association* helped both Japanese and Korean practitioners of biomedicine in colonial Korea publish their works while placing them in the broader scheme of Japanese supervision of biomedicine.

MATERIA MEDICA AND DISEASE PATTERNS: DESCRIPTIONS OF KOREAN ATTRIBUTES

Articles in the journals displayed three ways of depicting Korean attributes. The first was to investigate what was locally specific and deserving of being recorded for medical purposes. These articles imagined Korea as a unique region whose characteristics might reflect larger and more general topics. For instance, a survey on the materia medica of Korea was planned as a multivolume project to complete the entire body of knowledge about materia medica in East Asia.[42] Collecting data, analyzing the ingredients, and figuring out the medical efficacy of herbs had been of interest to biomedical experts since the early twentieth century.[43] The two journals differed in their approaches. Unlike other biomedical research, the investigation of Korean materia medica in *Medical World of Manchuria and Korea* legitimized the use of a group of Japanese scholars' in-depth knowledge of canonical texts of traditional medicine. Collecting terminology from all the classical texts of materia medica published in Korea and correcting different names of local botanicals became a major contribution of Japanese research.[44] For cataloging, different "kinds and grades" of Korean materia medica were analyzed, and the distinctions between "the orthodox and substitute items" and "different and fake items" were detailed. Furthermore, different names of Korean materia medica were traced as they appeared in traditional Korean and Chinese medical texts. In contrast, articles in the *Journal of the Chōsen Medical Association* centered on physicochemical characteristics of herbs, such as Korean ginseng.[45] Whether it was a philological inquiry into Korean texts of materia medica or a biomedical investigation of ginseng's medical efficacy, the "Chosŏn" category was used due to the Japanese conceptualization of the Korean locality.

The second method presented in these articles for researching Korean attributes was to describe the Korean environment in relation to certain patterns of infections. Triggered by an interest in a specific disease or a medical administration, the concept of "Chosŏn" Korea was employed to show how a disease or a medical condition was manifested in the bodies of Koreans.[46] As early as 1916, a "department of studying epidemics and endemic diseases" was established in the Government-General Hospital. This department aimed to examine Korean folk customs and their relationship to local diseases. With the support of the government-general, researchers were instructed to "examine Chosŏn's folk customs and its relationship with epidemics and to research endemic diseases prevalent in Chosŏn."[47] The hospitals founded by the Japanese, such as the Government-General Hospital in Seoul and the provincial hospitals, provided institutional settings to supervise hygiene and identify locally pervasive manifestations of endemic diseases.[48]

To some Japanese doctors, malaria and pulmonary distoma were understood to be the most serious diseases in Korea. Western-trained bacteriologists and zoologists were sent to examine insects to define the path of endemic diseases. By producing serum treatment for cholera and reporting the outbreaks of dysentery and scarlet fever, the Japanese attempted to scientifically analyze diseases common in Korea.[49] "Regarding Malaria in Chosŏn," "A Statistical Observation on Children's Typhus in Seoul," "An Epidemiological Observation on Scarlet Fever in Chosŏn," and "A Study on the Tetanus Bacillus in Korean Soil" are some of the research studies that aimed to report the situation.[50] Slight differences were found, but they mostly shared a common framework: individual studies clarified their sample size; used statistical methods; illustrated the results in tables and graphs; defined variables such as age, sex, educational background, progression of symptoms, and applied therapies; and detailed the relationship between those variables and the development of the disease. The conclusions were predictable, summarizing what the researcher found from statistical analysis, and often including the impact of Korean customs on the specific disease.

Research produced from a limited number of samples was made to represent Chosŏn Korea in its entirety. In "Regarding Malaria in Chosŏn," for instance, a Japanese specialist in hygiene reported on the serum test held at common (elementary) schools in Kangwŏn province, in central-eastern Korea. What this research actually accomplished was to calculate the percentage of malaria carriers among 116 students of one common school and 325 students of the other common school.[51] The age of students in this sample ranged from nine to twenty-two. The statistical method used in this research was a simple calculation of the ratio between carrier and the entire population surveyed. The sample size was small. Even so, this piece of research was able to represent Chosŏn Korea as a whole because the survey was linked to previous research that had reported the malaria situation of the entire Kangwŏn area at the district (*kun*) level. Based on these accumulated reports and cross-examination, the researcher was able to represent his survey as a general description of malaria in Korea.

Here, like "malaria in Manchuria" or "dysentery in Japan," the category of "Chosŏn Korea" primarily served as a regional unit in investigating the nature of diseases within the Japanese empire. "Chosŏn" as an ethnic unit of analysis became salient as it was connected to the Japanese construction of biomedical knowledge in East Asia. As a whole, Chosŏn Korean–bound research demonstrated the efficiency of the Japanese administration of public health and provided a framework for the Korean researchers of biomedicine to follow. Certainly, poor sanitary conditions, outbreaks of dysentery, and malaria were not uniquely Korean phenomena. Yet harmful diseases in the bodies of the colonized had to be carefully investigated and controlled, and reports on parasitic worms, malaria, distoma hepaticum, and syphilis required a range of administrative tactics. Statistical reports had to be carefully surveyed and accumulated for more than ten years to establish a more scientific approach.[52] The provincial hospitals and sanitary administration enabled Japanese medical agents to access the detailed conditions of a disease. The credibility of research would increase as more data, usually tens of thousands samples, were collected.[53]

Detailing disease patterns in Chosŏn Korea, researchers constantly compared Japanese and Koreans. For instance, a piece of research on scarlet fever in Korea, which measured average mortality in forty-seven major cities over twelve years, sorted the Japanese and the Koreans into two separate categories.[54] Statistical analysis on uterine cancer was performed by separately comparing Japanese and Korean cases.[55] Typhus in children in Seoul was statistically examined, addressing region, sex, age, and seasonal variables.[56] Along with these variables, the differences between Japanese and Korean inhabitants composed the major narrative structure of the report. The comparative and quantitative approach constructed a region or a people as an isolated category that had a clear and self-evident boundary.[57]

The comparative structure did not always support racial or ethnic prejudices, although Korean backwardness was unavoidably disclosed. A good example is the article on uterine cancer, which examined Korean and Japanese women in terms of cancer occurrence rates and followed treatments. The study analyzed outpatients who consulted the Gynecology Department of the Government-General Hospital between 1911 and 1924.[58] Various factors (cancer patients' age, number of past pregnancies, sterility, and career, as well as the time until cancer was detected, the patients' symptoms, the exact site of cancer detected, whether operation was possible, and whether the cancer was recurrent) were investigated. Ratios for these variables were then calculated between groups. Again, a simple arithmetic technique was applied, yet due to the Japanese control of data collection and organization, which reported the entire number of outpatients (26,999 for eleven years) and their clinical information, the individual article gained meaningful ground for articulating its own findings. The overall tone of comparison reflected the ideals of science at the time: being objective, quantitative, and concise. The research showed both similarities and differences between the two groups; the cancer was most frequently found in Japanese women in their thirties and in Korean women in their forties or fifties. In both cases, the cancer was most frequently detected in the cervix; the ratio of operable cases was slightly higher in Japanese cases, but recurrence was also higher in Japanese cases. The quantitative data, however, revealed that uterine cancer was a more serious problem among Korean women, as it was demonstrated to have a higher occurrence rate (Japanese, 0.62 percent; Korean, 2.09 percent) and more advanced symptoms among Korean women who first visited the hospital for a medical consult. Researchers ascribed this to the overall inconvenience of public transportation and limited knowledge of biomedicine in Korea.

In sum, the assumed idiosyncrasy of Korean soil and the envisioning of the region as a unit of disease manifestation solidified the category of "Chosŏn" within the Japanese colonial management of biomedicine. A set of bifurcated comparisons, combined with a mathematical analysis, established "Chosŏn" as an indispensable label of scientific ethnography. In building up the two aforementioned patterns, Japanese biometric research rigorously used the category of "Chosŏn," elaborating complicated meanings of comparison and objectification during the 1930s.

BIOMETRICS: DESCRIPTIONS
OF KOREAN ATTRIBUTES

In addition to the two aforementioned methods, a third way of depicting Chosŏn Korea was found in Japanese biometrics. During the 1930s, Japanese medical researchers rigorously quantified bodies of Koreans, detailing general physique, size of skulls, or blood types.[59] A hypothesized curve of normal distribution was frequently employed as a representational tool. The growing zeal for rationalizing racial and ethnic differences spurred the measurement of Korean juvenile delinquents' foot sizes to investigate their relationship with height, yet more subtly to look for a physical type that characterized the juvenile criminal.[60] The physiques of Korean students who entered the professional schools were examined to detect the relationships among nationality, intelligence, and physique.[61] The meticulous biometric gaze even traced the different shapes of the crowns of Korean heads, the diverse appearance of Korean juvenile prisoners' tattoos, Korean female students' menarche, and the calcium component of healthy Koreans' blood.[62] Beyond Korea, the racially specific category was applied to Manchuria in the late 1930s; thus, articles that aimed to fully report Manchurian and Mongolian physiques and endemic diseases were often found in the journal issues of the late 1930s and early 1940s.[63]

The full-fledged biometrics aimed at Koreans was associated with the changing mode of race discourses in Japan. As Tessa Morris-Suzuki persuasively demonstrates, ideas about "race" became more diverse as Japanese intellectuals more fully employed foreign theories of scientific racism and Western critiques of racial science during the 1930s. The German Nazis' idea of "racial purity" and Julian Huxley's and A. C. Haddon's criticism of scientific racism fueled Japanese disputes on race.[64] In spite of the multiple currents of ideas, a convergence toward the notion of a "single line of human progress" undeniably persisted.[65] For instance, Kada Tetsuji (1895–1964) overtly rejected the idea of scientific racism, yet he firmly believed in the progress of society, which would be led by a racially homogeneous and ethnically refurbished nation-state. Morris-Suzuki succinctly points out that essentialized race, whether or not it was fully justified by science, was linked to the blood-based differentiation of human beings; hence, it was easily transformed into an idea that would support Japanese imperial expansion, acts of segregation and discrimination, and a sense of national superiority.

Given the evolution of the race discourses in Japan, it is noteworthy that the Japanese biometrics about Korean bodies in the 1930s diverged from that of the 1910s with an increased number of articles on the subject, a diversified analytical scope, and a nuanced portrayal of racial and ethnic differences. The Japanese quantification of Koreanness changed over time, reflecting the malleable and often confusing nature of scientizing the indigenous "Chosŏn." Kubo Takeshi's anatomical studies about Korean bodies provide a locus to consider the (dis)continued patterns of Japanese biometrics, as well as its impact on Korean doctors of biomedicine.

Kubo Takeshi in the 1910s. Kubo Takeshi published multivolume reports about "Racial and Anatomical Research on Chosŏn Koreans" in the *Journal of the Chōsen Medical Association* between 1917 and 1922.[66] Born in 1879, he earned his medical license in 1898, and established his expertise in anatomy through his training in Tokyo, Seoul, and Manchuria. Kubo Takeshi's 718-page doctoral dissertation about the physical anthropology of Koreans was rendered in German and submitted to Tokyo Imperial University in 1913. In this dissertation, he meticulously measured a grand total of 3,425 living Koreans, and his analysis was featured by the *Journal of the Anthropological Society of Tokyo* in 1914.[67]

Building on his initial success, Kubo Takeshi aimed to include samples from available Korean cadavers. As he repositioned himself as a professor of anatomy at the newly opened Professional School of Medicine in Keijō in 1916,[68] Kubo Takeshi was able to access the corpses of executed prisoners and the unclaimed dead bodies of Korean patients. His sample was composed of forty males and six females. He numbered each dead body, specifying the institutional affiliation, age of death, place of birth, and cause of death.

Kubo Takeshi's research framework displays his zeal to typify a race. With an archetype in mind, he wanted to define a normal Korean body. Due to the limitation of available corpses, according to Kubo Takeshi, the existing anatomical research had mostly focused on abnormality caused by illnesses. Criticizing this limitation of earlier studies, Kubo Takeshi aimed to scientifically define the prototype of Koreans. Only nine out of forty-six corpses were obtained from persecution; however, by showcasing the bodies that were free from any serious diseases, Kubo Takeshi hoped to depict his research as about "normal" Koreans.[69] He selected a criminal's body and then depicted it as representative of Koreans in general. Cadaver number seven, labeled K. T., an imprisoned forty-one-year-old male from Kyŏng-gi province, 154 cm tall and weighing forty-seven kilograms, became the significant sample of the Korean.

Kubo Takeshi's idealization of a "Korean" body required an equally reductionist Japanese counterpart. It was nearly impossible to identify the traits of "Korean" bodies without employing countless comparisons with "Japanese" ones. Like two sides of a coin, the weight and length of Korean criminals' muscles only gained scientific significance through the interpretative comparisons with the Japanese counterparts.

It is interesting that Kubo Takeshi actually complained about the lack of scientific research about Japanese bodies, which hampered the advanced understanding of Korean samples. Kubo Takeshi expressed his regret that the existing statistical research about Japanese body parts—such as weight and length of muscles, internal organs, hair, and skin—was quite rare and partial. To tentatively overcome this shortcoming, he relied on the corpses he had measured in Japan a year earlier. All together, his Japanese samples amounted to eleven male, two female, and two child corpses—far fewer than the Korean samples. Kubo Takeshi selected a forty-year-old Japanese criminal, 156.5 cm tall and weighing forty-eight kilograms, as representative of the "Japanese." Given the insufficient number of corpses and lack of scientific consensus about the "Japanese" body, Kubo Takeshi imposed a reductionist gaze on both Koreans and Japanese.

Kubo Takeshi continued to express his regret about the lack of Japanese samples for comparison in his detailed reports of the 1910s. The 342-page report was rendered in Japanese and organized by the muscles of different body parts: "muscles of back and front" sets the major organizational structure. The back muscles were divided into two categories, shallow and deep; under these two divisions, Kubo Takeshi measured twenty-two kinds of muscles. The "front body" part detailed subdivisions of head, neck, chest, abdomen, sacrum, and the four limbs. Kubo Takeshi's writing was clear and concise.[70] Using different sizes of tables, he shrewdly reported the Japanese and Korean differences. The median of his samples was calculated, but his Japanese sample usually implies the one representative corpse, whereas the Korean counterpart was three to seven corpses. For instance, when Kubo Takeshi specified muscles around the mouth, he compared only one Japanese to a Korean, then to the median of three Korean samples. In a similar vein, one table featured the upper muscles of the hyoid bone, contrasting only the one Japanese representative to a

Korean counterpart.[71] The limited number of samples was evident in many tables throughout his report.[72]

Kubo Takeshi enhanced his comparative approach by adding the "European" (*Ōushūjin*) category to the existing Japanese and Korean ones. He did not explicitly define the exact scope of "Europeans," yet the use of the category enabled him to rely on German references, thereby legitimating his reductionist framework. For instance, he pointed out that "Mr. Vierordt's Daten und Tabellen "(Mr. Vierordt's data and tables) analyzed a single case of a criminal in generalizing Europeans' muscle weights,[73] and Kubo Takeshi applied this research convention to his own analysis.

Interestingly, the three-way comparison enabled Kubo Takeshi to highlight the similarities between Koreans and Japanese in contrast to the European case. In comparing muscle heights and weights, the gaps between Koreans and Japanese were less explicit than those between Koreans and "Europeans." The similarities between Koreans and Japanese, therefore, were persuasively accentuated through the introduction of the European category.

Kubo Takeshi has been notorious for imposing racial prejudice against Koreans. His biometrics in the 1910s surely exposes the limitation of the "scientific" portrayal of race. His artificial research design, combined with insufficient data, failed to persuade even his contemporaries. A careful reading of his polemics exhibits the degree to which his typifying efforts applied to the bodies of Japanese as well. Kubo Takeshi longed for not only the "Korean" archetype but also the scientifically articulated "Japanese" pattern. In addition, his detailed quantification, mediated through a three-way comparison, underscored the similarity between Japanese and Koreans as early as the 1910s. Kubo Takeshi's "scientific" investigation of race, therefore, exposes components for both inexplicable differences and intrinsic overlaps between Koreans and Japanese.[74]

Kubo Takeshi Revisited in the 1930s. The Japanese biometrics during the 1930s primarily demonstrates efforts to diverge from Kubo Takeshi's earlier framework in the 1910s. The research directed by Ueda Tsunekichi (1887–1966) and Imamura Yutaka (1896–1971), for instance, aimed to upgrade their protocols by avoiding Kubo Takeshi's subjective and reductionist typology of race.[75] Both scholars led the Department of Anatomy at the Keijō Imperial University in Seoul, supervising a series of research studies about Korean physique.[76] Both were trained in Japan, and then in 1926 became affiliated with the newly established Medical College of

Keijō Imperial University as professors of anatomy. From the beginning of their careers in Korea, Ueda Tsunekichi and Imamura Yutaka collected Korean skulls, honing their technique of biometrics. Between 1934 and 1936, their physico-anthropological study on Chosŏn Koreans and neighboring ethnic groups was supported by the Hattori Hōkōkai Foundation; its total cost amounted to ¥2,000.[77]

A ninety-page report titled "A Physical Anthropology Study on Chosŏn Koreans," supervised by Ueda Tsunekichi and Imamura Yutaka, aimed to critically supersede Kubo Takeshi's earlier research. Referring to Kubo Takeshi's 1913 doctoral dissertation, the authors lauded his unprecedented achievement in synthesizing many samples and his rigorous elaboration of measuring categories. Simultaneously they criticized his biased methodology: Kubo Takeshi relied heavily on soldiers for physique measurements, as they were easily accessed, and selected his samples mostly from the Kyŏnggi province, near Seoul. More to the point, his statistical methodology was outmoded. According to the 1930s researchers, Kubo Takeshi's reductionist approach to demonstrating "purely" Korean traits was not fully "scientific."

Instead, the new researchers aimed to polish their statistical techniques while diversifying their sample selection. More input was given to quantitative analysis, thus it is no surprise to see the report was made up of almost one hundred tables. The first half of the team report was published in 1934, displaying eighty-seven tables and three diagrams. The second half, published in the same year, included 778 new samples from four southern provinces—Hwanghae, Kyŏnggi, Kangwŏn, and Ch'ungbuk—and the forty pages exhibited a variety of tables showing the arithmetic mean and standard errors. Except for the beginning, where the overall research goal was manifested, the authors' interpretative statements were rarely presented. This contrasts to Kubo Takeshi's verbosity, as he frequently disclosed his subjective assumptions about racial differences. The authors in the 1930s also compared Koreans with Japanese; hence, northern Koreans' relatively tall height was contrasted with the average Japanese's shorter stature. Unlike Kubo Takeshi, the later researchers neither carried out a full-fledged comparison nor used the contrast of the two groups as the major narrative structure.

The noticeable evolution of the Japanese measurement of the Korean physique lies less in the scientific demonstration of purely Korean attributes than in the process of making the "Korean" race. Not the unchanging ground but the evolutionary procedure of race became a main agenda. Thus, the research of the 1930s paid special attention to northern Koreans, as their physical traits, when compared with neighboring ethnic groups, might disclose the ancient route of Korean migration into the peninsula. In four northern provinces, including northern and southern Hamgŏng and northern and southern P'yŏngan, the researchers collected 639 samples, then detailed the measurements of height, various parts of the torso, and limbs. The examinees' sex, age, regional, and vocational divergences were considered, and the arithmetic mean, standard deviation, variation coefficient, and standard errors were calculated. Although the researchers concluded only by highlighting the regional divergences in Korea, the underlying aim was to put the divergent traits of Koreans on a larger map of racial constitution in East Asia.

Whereas Ueda Tsunekichi and Imamura Yutaka's research represents the efforts during the 1930s to shift away from Kubo Takeshi's biometrics, Kubo Takeshi's legacy continued in the 1930s. Itaru Shibata's (n.d.) measuring of the brain weights of Koreans fleshed out Kubo Takeshi's discussion about "similarities" between Koreans and Japanese, reflecting similar research carried out in the 1930s. Affiliated with the Department of Anatomy at the Medical College of Keijō Imperial University, Itaru Shibata aimed to demonstrate the meaningful relationship between brain weight and other variables, such as overall body weight, age, and sex. The lack of Korean-related research motivated him to examine corpses. Through his affiliation with the imperial college, he was able to gain access to 136 male and 17 female brains.[78]

Itaru Shibata followed the conventions of other biometric research of his time. The arithmetic averages of his sample brains were analyzed according to sex and age, and their relevance to height was considered. Throughout the article, Itaru Shibata highlighted the meaningful similitude between Japanese and Korean brain weights and its relevance to height. He asserted, "The periods of increase and decrease of the brain weight in the Koreans, therefore, seem similar to those in the Japanese."[79] Generally speaking, the Korean data confirmed the existing theory of correlation: "The larger the body length, the heavier the brain weight."[80] Considering the general principles of physiology, Itaru Shibata concluded that "it does not appear, nevertheless, that the brain weight relative to the body weight in the Koreans shows any distinct differences from that of the Japanese."[81] Overall, "the brain weight of the male Korean is in the same category with that of the Japanese."[82]

Two existing scholarly discourses seemingly affected Itaru Shibata's emphasis on similarity. First, as Itaru Shibata mentioned, Kubo Takeshi's biometrics had already detailed manifold overlaps between Japanese and

Koreans, and Itaru Shibata was building on this scholarship. His focus on "similitude" did not suddenly emerge in 1936. Rather, he learned from Kubo Takeshi's earlier research, extended the number of samples, and refined the comparative angle. As a result, Itaru Shibata competently concluded that Japanese and Koreans revealed no significant differences in terms of the physiological relevance of brain weights.

Second, the accumulated research about Japanese brain weight by many authors made it less effective to propose racial purity, such as the mutually exclusive abstraction of "Japanese" or "Korean." In Kubo Takeshi's earlier research, the arithmetic average of male brain weight of Koreans was 1,353.2 mg, whereas Itaru Shibata's article calculated it as 1,369.5 mg. The difference between these results was 16.3 mg. On the contrary, the Japanese data by then showed a wide divergence. Whereas a researcher reported the smallest mean average of 1,348 mg by measuring twenty-four male cadavers in 1892, another scholar reported the heaviest mean average, 1,406 mg, in 1930. The difference here exceeds 58 mg. The accumulated Japanese research hardly showed any consensus about the Japanese brain weight. The difference among Japanese samples exceeded the gap between the Japanese and Korean data. The existing and disagreeing Japanese data possibly made Itaru Shibata avoid the overly reductionist comparison between racial biometrics. Instead, he focused on the relationship between the brain weight and height, or body weight, then concluded that Koreans' brain weight belonged to a similar physiological category as that of the Japanese.

The existing scholarship argues that the emphasis on similitude was caused by the enhanced Japanese assimilation policy, militarization, and mobilization of Koreans in the 1930s and the early 1940s.[83] Given the overall political atmosphere of the time, however, more thought should be given to complicated motives and methodologies in order to fully analyze the interplay between political ideology and scientific research framework. A nuanced understanding defies a mechanical and forceful imposition of a policy on scientific terminologies and theories. The idea of similarity between Japanese and Koreans did not suddenly emerge in the 1930s. From the 1890s, Japanese measurement of the Korean body had honed its analytical technique, augmenting the number of samples and controlling manifold variables such as regional divergences. More to the point, the essentialized objectification of "Koreans" required the similarly reductionist categorization of "Japanese," and the mechanical contrast between two groups sounded less persuasive, given the growth and articulation of biometrics in the 1930s.

Viewed as a whole, the Japanese biometrics came to display an explicit yet confusing conceptualization of Korean indigenousness. The difference between Japanese and Koreans was assumed to be self-evident, as Kubo Takeshi's research well demonstrates. Without essentializing the mutually exclusive realm of Chosŏn "Koreans" and "Japanese," Kubo Takeshi's ambitious project could not be properly carried out. The acceptance of an innate difference among races or ethnic groups was a prerequisite for the rigorous quantification of Korean bodies. Yet as race discourses evolved over time and biometric research was accumulated, the techniques of conceptualizing difference highlighted dissimilar points.

In sum, Japanese used "Chosŏn" Korea as a unit of analysis in medical discourse to look at local environments and their medicinal products, to explore the geography of diseases like malaria, to control, and to find the racial and ethnic characteristics of Korean bodies. The biomedically constructed Korean body offered salient contrast and comparison with the Japanese self, yet confusion and revision accompanied Japanese practice of biometrics. In the Japanese biomedical narratives, Chosŏn Korea as a locality became a project, not a self-evident fact, which reveals patterns of scientific knowledge in identifying, translating, and circulating the local within the geographically and culturally growing Japanese empire in the 1930s.

The Japanese biomedical description surely required a variety of institutional conditions. Primarily, the Japanese research on Korean bodies was accumulated as a whole body of knowledge. A statistical survey on a specific issue was therefore validated within an already well-established collection of knowledge. Second, the production of biomedical knowledge took place in the institutional setting of the Japanese-led hospitals and school of medicine. To be an authentic knowledge producer, one needed to learn the protocol of research. Every article was supposed to follow the structure, the same type of tables, and rhetorical conventions of Japanese scholarship.

The Japanese analysis of the Korean body significantly shaped the way Korean doctors of biomedicine framed their research. Newly emerging Korean professionals of biomedicine emulated similar scholarly conventions without questioning the underlying assumptions or the scientific credibility of the research design. The following section examines the nature and problems of Korean-produced biomedical knowledge about the bodies of Koreans.

The Views of Korean Doctors of Biomedicine

WAS THE LABEL OF "CHOSŎN" VALID?

The number of Korean doctors trained in biomedicine under Japanese colonialism amounts to approximately 4,000. The number increased from less than 100 around 1910 to 2,600 in 1943.[84] As Korean doctors grew in number, so did the longing for an initiative. Korean doctors and dentists with biomedical training founded the indigenous Association of Korean Doctors (Chosŏn ŭisa hyŏphoe) in 1930. Membership was limited only to Koreans (*Chosŏnin*) who were able to pay membership fees. The association wanted to "exchange knowledge in medicine, to promote close relationship among members, and to disseminate and upgrade ideas of hygiene."[85] Seoul became the major hub of the association, and the doctors who were affiliated with Severance Union Medical College played a leading role in publishing the society's journal, *Chosŏn ŭibo* (Korean medical journal). Rendered in Korean vernacular, the journal was published between 1930 and 1937 in seven volumes and twenty-seven issues. Although short lived, these publications in vernacular language displayed various concerns and strategies in applying biomedicine to specific issues of Korean health and diseases.

Not all articles in the *Korean Medical Journal* investigated indigenous attributes. However, the researchers joined in elaborating "Chosŏn," showing (dis)similarities to the Japanese counterpart.[86] The noticeable difference lay in the lack of interest displayed for investigating native products, or materia medica, in Korean soil. Only one article during the seven years attempted to demonstrate the efficacy of ginseng. Indexing native herbs and scientifically examining their therapeutic potential was not a major agenda among Korean doctors of biomedicine. Only one article positively conjectured the therapeutic potential of traditional medicine. Yi Se-gyu (1905–86), who was affiliated with the Department of Pharmacology at Severance Union Medical College, advocated for the validity of Eastern medicine.[87] He referred to the latest Japanese biomedical research on the role of stimuli caused by moxibustion, then speculated about the possible scientific grounds of acupuncture and moxibustion. Except for these two articles, no Korean research further examined native herbs or the indigenous heritage of medicine, which contrasted with the visible upturn in these topics in Japanese research.

Notwithstanding this minor disparity, Koreans emulated the Japanese use of "Chosŏn" extensively. In general, the *Korean Medical Journal* showed no significant departure from the Japanese publication *Journal of the Chōsen Medical Association* in terms of editorial structure, covered topics, and research methodologies. Each installment of the Korean journal similarly issued a couple of research articles, delivered the latest clinical reports, served as a newsletter for its members, and included ads from major pharmaceutical companies and apothecaries. Moreover, Japanese or German references were preferred to English ones, and Korean researchers prioritized quantitative analyses with arithmetic means and standard variation, frequently displaying tables and diagrams.[88]

Following the Japanese model, the Korean doctors hoped to scientize Korean characteristics. Korean researchers used the "Chosŏn" label, imagining "average" Koreans or "normal" Korean attributes. Under the competing atmosphere of internationalism in medical science, Koreans unquestioningly portrayed the homogenized categories of Chosŏn Korea or Koreans, then aimed to circulate the Chosŏn label in line with the German or Japanese units of analysis. In this framework, the magnesium content of healthy Chosŏn Koreans' blood was measured.[89] The impact of pungent spices (which characterize Korean dining in general) on the secretion of gastric juice was examined.[90]

Not surprisingly, the normalizing framework of "Chosŏn Koreans" nurtured comparison and contrast. Yi Yŏng-ch'un (1903–80) and Ch'oe Chae-yu (1906–93), for instance, examined the blood sugar content of "healthy Korean adults."[91] Challenged by Japanese, German, and Chinese data, the authors measured thirty-four men and fifty women who were considered healthy.[92] Another article about the magnesium content of healthy Chosŏn Koreans' blood revealed a similar research design. The author examined fifty Koreans aged ten to forty-five, then compared the amount of magnesium in their blood (a maximum of 1.77 mg and a minimum of 1.32 mg) to the already-known data of Americans and Japanese.[93]

Interestingly, the "Chosŏn" category presumably applied to demonstrate Korean characteristics often failed to fully prove its ethnic distinction. According to Yi Yŏng-ch'un and Ch'oe Chae-yu, blood sugar content measured by the Hagedorn-Jensen method was an average 0.107 percent among Koreans, which was not significantly different from the Japanese average of 0.110 percent. The Koreans' blood sugar content was expected to be higher than that of Chinese or Europeans because Koreans were known to consume more carbohydrates and less fat. However, the average of Korean blood sugar was not different enough to reflect any specific ethnic features.[94] Likewise, the magnesium content of healthy Chosŏn Koreans' blood, according to the researcher, exhibited a noticeable difference compared with Americans but not with Japanese.[95]

Korean researchers aptly used the analytical unit of "Chosŏn," yet their conclusions negated the ethnic peculiarities that were initially assumed. A series of biometrics most vividly demonstrates this pattern of interpretation. Emulating the Japanese convention, Koreans aimed to quantify and compare the size of the pelvises, Korean infants' nail lines, girls' age at menarche, and blood types. In these pieces of research, "Chosŏn" was conjured up as a term, a kind, and an entity alluding to the physiological and cultural distinctiveness of Koreans. However, the conclusion did not fully support the assumed bodily distinction of "Chosŏn" Koreans. In a sense, the Korean doctors' appropriation of "Chosŏn" in medical writings was a response to the Japanese focus on biometrics and racial science. However, the Korean employment of the unit of "Chosŏn" revealed limits and bewilderment, which eventually made the ethnic differentiation unstable.

An example comes from Paek T'ae-sŏng's (1904–63) measurement of Koreans' pelvises (os coxae). Unlike German or Japanese research, which was exclusively focused on their own racial features, Paek T'ae-sŏng found that the research on Korean cases had been largely neglected. Aiming to make Korean bodily traits visible, he examined forty-three corpses, but the measurement of pelvic angles did not reveal any apparently Korean attributes. Paek T'ae-sŏng concluded that Koreans' pelvises were not actually different from those of the Japanese.[96]

In a similar vein, Yi Sŏn-gŭn (1900–66) examined physiological growth of infants' nail lines and compared data between Japan and Korea. He examined 126 infants aged 19 to 144 days. He assumed the physical growth of healthy and normal infants correlated with the growth of the nail line, which has an implication for forensic medicine. However, he failed to demonstrate any meaningful differences in elaborating this idea. He added that in terms of the "physiological nail line, within the scope of my research, Koreans and Japanese had no differences."[97]

Ch'oe Tong (1896–1973) examined parasitic worms in Chosŏn Korea, yet his research ended up revealing more resemblance than difference between Japan and Korea.[98] Ch'oe Tong explained that although Korea was notoriously known as the "paradise of parasite," and therefore assumed to be different from the Japanese case, this was only true in terms of intensity; compared with Japan, Ch'oe Tong found, the kinds and the range of distribution of parasitic worms were not very different.

Certainly, not all research was inclined to highlight the similitude between Japanese and Koreans. Yun Ch'i-wang's (1895–1982) research about the age of Korean girls' menarche counted 9,429 cases, then presented the mean value of the age distribution without asserting any similarity between Koreans and Japanese.[99] However, it is not an exaggeration to say that the Korean biometrics in general underscored Korean similarities to the Japanese during the 1930s.

As Japanese racial discourse evolved, so did the Korean interpretation of indigenous characteristics. To map blood types, Ch'oe Tong examined 1,300 Koreans, including patients, students, and nurses, who were affiliated with the Department of Medical Jurisprudence at Severance Union Medical College.[100] The incidence of four blood types (A, B, AB, and O) was determined and the relative percentages calculated to examine the relationships among blood types and regional differences, one's domicile of origin, and surname differences. The survey resulted in a racial index number of 1.0, which placed Koreans in the spectrum of racial differences that ranged from 9.6, the highest number for Native Americans, and 0.6, the lowest number for Indian people. Ch'oe Tong cautioned that the racial index did not measure racial superiority.

Whereas the index numbers enabled Koreans to be positioned in the international range of racial differences, divergences within the nation were analyzed to infer a genetic relationship between those places where the same blood types were concentrated. This typifying and connecting of regions resulted in a few vertical lines that penetrated Korea from the north to the south. Based on this connection of dominant blood types, Ch'oe Tong theorized that the path of the Korean race migration stretched from Manchuria and Mongolia to the Korean peninsula. Although his sample did not fully demonstrate his theory, he did not hesitate to visualize this ancient path according to the mapped blood types. Instead of highlighting racial purity, he explored more of the mutation, migration, and modification of races, which resonated with the Japanese research trend in the 1930s.[101]

Korean medical studies in the 1930s seem unscientific today. Yet they did not merely display the lack of qualified research skills. Instead of evaluating the Korean research as ill equipped, inconsistent, and groundless, I view the hesitant and tentative usage of "Chosŏn" as a response to the unstable and marginalized conditions to which the Korean doctors of biomedicine had to adapt. First and foremost, the emergence of nationally and racially bound units of analysis motivated Korean doctors, who felt they were left behind in modern medicine, to use the ethnic category in developing scientific research more easily and concretely. Taking the Japanese and German research as examples, Korean doctors hoped to establish the explicit "Chosŏn" category. Setting a "Korean" component, therefore, fit well with the scientific convention of the 1930s. More to the point, beneath the label of "Chosŏn," we sense a growing desire among Koreans to be agents—not mere objects—of knowledge production. The fledgling Korean doctors of biomedicine attempted to take the initiative in investigating the bodies of Koreans.

Second, it is worth noting that the scientifically examined Korean body did not always succeed in concretizing racial or ethnic distinctions. The "Chosŏn" category gave Korean doctors of biomedicine an opportunity to depict indigenous peculiarities by using the universal grammar of science. However, the marginal position of Korean biomedical doctors caused the confusing nature of Korean distinction, which was imposed and then eventually aborted. It should also be noted that Korean doctors of biomedicine did not fully control the collection and organization of data. Unlike Japanese researchers, who were able to access official data in every province and did not hesitate to draw a nationwide map of disease and health, Koreans usually studied with at most tens or hundreds of pieces of data that were collected at a private institutional level. Most Korean research did not project any decade-long surveys. Under the Japanese empire's maintenance of biomedical knowledge, the circulation of "Chosŏn" as an analytical unit was carried out by Koreans for an initiative in vernacular. However, the selected category failed to signify a meaningful cultural or ethnic distinction in medical research.

WHO WAS THE CHOSŎNIN DOCTOR OF BIOMEDICINE?

Paralleling the unstable nature of the Chosŏn category, we find that the Korean biomedical doctor maneuvered identity as both a cosmopolitan professional and a colonial subject. Ming-Cheng M. Lo casts light on the Korean experience by analyzing the manifestos and political activism of biomedicine-trained Taiwanese doctors under Japanese colonialism. Lo persuasively demonstrates the conflicting interplay between ethnic distinctions and liberal professionalism. Taiwanese doctors shifted away from identifying as national physicians, which had situated medical knowledge within specific sociocultural circumstances, and moved toward a liberal sphere of modern medical professionals as the Japanese regime attracted them and placed them more firmly within the medical cosmopolitanism and imperialism in the 1930s.[102] Lo's interpretation illuminates the experience of the doctors' Korean counterparts, who also struggled at the intersection of Chosŏn Korea's particularism and biomedicine's cosmopolitanism.

Paek In-je (1899–?), one of the most successful surgeons in colonial Korea, exemplifies the manifold constraints encountered in engaging Korean specificities with the culture and status of biomedicine.[103] Born in Chŏngju, a major center of enlightenment ideas in the northern province of P'yŏngan, Paek In-je was nurtured by growing nationalism and anti-Japanese activism. In his third year in the Professional School of Medicine in Keijō, he participated in the March First Movement and consequently suffered in prison for ten months. Although Paek In-je came back to school and graduated first in his class, he had to work as an apprentice without salary at the Government-General Hospital for two more years to obtain his license. After his participation in nationalist activism, he seemed to seriously consider his future as a doctor of biomedicine. When he returned to his studies, he decided to prioritize professionalism over any kind of armed anti-Japanese movement.[104]

Paek In-je entered the Professional School of Medicine in Keijō right after it was upgraded from a training school. Among the seventy-nine new students in 1916, Koreans amounted to fifty-four and Japanese twenty-five. By 1945, the school produced approximately 823 Korean and 935 Japanese graduates since attaining an upgrade.[105] It is no wonder that ethnic conflict between Korean and Japanese students was reported. As discussed earlier, Korean students felt offended when anatomy professor Kubo Takeshi lectured on racial differences between Koreans and Japanese. Despite these documented clashes, however, students reported being mostly satisfied with the quality of teaching at the school. All the textbooks were in Japanese, and it was required for students to master German to understand some medical terms translated from German into Japanese.[106]

Although Paek In-je had been inspired by nationalist activism, he had not openly expressed antagonism toward Japanese colonialism since the 1920s. He built quite a good relationship with Japanese teachers, and it continued to develop while he taught in the Professional School of Medicine in Keijō as an instructor for the Department of Surgery. In particular, Satō Gōzō (1880–1957) and Uemura Shunji (b. 1876) were known to support him during hard times. When Uemura went back to Japan in 1941, he transferred his clinic to Paek In-je.[107] This clinic became the predecessor of the present Paek Hospitals, which celebrates Paek's legacy.

Paek In-je joined the Association of Korean Doctors, creating a space for native initiatives in biomedicine. He gave public lectures aiming to enlighten Koreans on the matter of health management. He had personal relationships with Korean nationalists such as Philip Jason (Sŏ Chae-p'il, 1864–1951). Despite Paek In-je's interest in Korea's specific geopolitical situation, however, it is not difficult to see that he was drawn more to cosmopolitanism as he successfully built his career during the 1930s. He believed that mastering advanced knowledge, whether it originated from the West or was mediated by Japan, would ultimately resolve Chosŏn Korea's problems. Paek In-je assumed Western-style democracy and capitalism would eventually govern Korea. While he helped his daughter study English, he underscored the significance of learning it not just as a foreign language but also as a tool for further understanding Western civilization.

Paek In-je's travelogues in the late 1930s display his sensitivity to and willing acknowledgment of Japanese hegemony in East Asia. Because his hospital was managed successfully, he had opportunities to go abroad to view Western medical administration.[108] On his first trip he spent a few months in Europe in 1930, and on the second he traveled to Europe and the United States between November 1936 and January 1938. During this visit, Paek In-je visited the Mayo Clinic in Rochester, Minnesota, as well as the Johns Hopkins Hospital and Columbia University. He also stopped by to see well-known medical centers in Germany, England, France, Italy, Austria, Denmark, and Norway. Paek In-je's journey to the West was reported in Korean newspapers, and his account of the trip was published in the *Korean Daily* from January 13 to 27, 1937. For forty days, his journey continued, and his ship stopped at various ports in China, Hong Kong, Singapore, and India. Given the rarity of international travel for most Koreans, Paek In-je did not hesitate to fully describe exotic scenes and foreign customs, in addition to his impressions of the rapidly changing world. In particular, he addressed Japan's expanded market to South Asia and warships anchored at various ports in Southeast Asia, which demonstrated the empire's military prowess. Paek In-je also appreciated the well-mannered Japanese, who were the majority of passengers. Portraying the Japanese way of socializing, he said that they were the most cultured among the passengers, including Europeans and the Chinese. Paek In-je was the only Korean on board the ship.[109]

This brief story about Paek In-je can hardly exemplify a general trajectory of biomedicine-trained Korean doctors during the 1930s. Even compared to his colleagues, Paek In-je's career, status, and experience were extraordinary. He came from an old, elite family, and his education and biomedicine were two fields in which he experienced upward mobility. By the end of the 1930s, his way of life was far removed from that of most Koreans, both culturally and economically. Given the disparity, however, his brief biography exemplifies a shared ground among colonial elites of biomedicine, who experienced the conflicting ideals between liberal professionalism and nationalist particularism. Paek In-je and the aforementioned authors who actively used the "Chosŏn" category were drawn to engaging Korea's specific circumstances with their advanced knowledge of biomedicine. The scientizing framework of the "Chosŏn" neither succeeded in always solidifying local characteristics in their profession nor led to any noteworthy innovation of health care in colonial Korea. Their privileged socioeconomic background showed the ever-growing gap between doctors of biomedicine and other fellow Koreans. Under the Japanese empire in the 1930s, Korean doctors of biomedicine accentuated and simultaneously trivialized the embodied ethnic or racial distinctions inasmuch as their Japanese counterparts exposed both a clear and simultaneously confusing differentiation of "Chosŏn."

Conclusion

At the global and local junctures of scientizing human specificities, "Chosŏn" emerged as prominent unit of analysis among medical professionals in the 1930s. By analyzing biomedical research papers, popular essays, and newspaper articles, this chapter shows that the idea of the Chosŏn's specificity in relation to health and diseases was not an intrinsic trait that would be discovered, but was an outcome of contesting discourses and compromising research methods. Medical explorations of the Chosŏn's particularity in the 1930s exhibited manifold interactions between people, knowledge, and institutions. Advocates of indigenous medical heritage aimed to borrow terms, images, and educational models from the newly emerging profession of biomedicine. Practitioners of traditional medicine rejected biomedicine's monopoly on universal authority, thereby hoping to provincialize Western medicine. However, in reality, advocates of traditional medicine admitted the superior status of biomedicine, and hence planned to scientize the millennium-old knowledge and conventions of their field. Japanese medics' growing interest in "Chosŏn" provided reliable resources for Koreans in arguing for the validity of traditional medicine. Similarly, the Japanese consolidation of "Chosŏn" in their rigorous cataloging of herbs, diseases, and physiques shaped the way newly established Korean professionals of biomedicine analyzed the indigenous attributes of health and illnesses. Although motivated by nationalist sensitivity and rendered in vernacular language, the Korean elaboration of "Chosŏn" often lost its initial validity, then ended up by highlighting the similar physiological nature of Japanese and Koreans. The increased circulation of the "Chosŏn" label in the 1930s testifies more to the term's instability than to its solidification.

Chapter Four

Lifesaving Water

Managing the Indigenous for Medical Advertisements

On September 1, 1911, a full-page advertisement of a Japanese Indan (Jintan in Japanese, Humane elixir) appealing to Korean consumers ran in *Maeil sinbo* (Daily News).[1] First manufactured by Morishita Hiroshi (1869–1943) in 1900 as a mouth refresher, Humane Elixir became one of the best-selling patent medicines in East Asia.[2] Consisting mostly of sugar and aromatic substances, it promised both physical and mental rejuvenation. Although its all-encompassing medical efficacy sounds outmoded today, the graphically fresh advertisement and its deliberately designed health advice call for further analysis.

A smiling lady in Western dress stands with open arms at the prow of a sailing boat, hinting at the voyage toward a modern world (Figure 1). In one hand she holds Humane Elixir, and in the other, its container, which was marketed as a promotional gift. The English-language caption urges readers to carry it all the time: "Whenever, however, and wherever you may be, you'd surely enjoy excellent health in taking Jintan as a task. Jintan is a wonderful tonic and the best mouth refresher."[3]

In addition to Korean, ads were also rendered in Japanese and Chinese, revealing the multilinguistic environment of ad production in early 1910s Korea. It is likely that a Korean agent hired at the pro-Japanese newspaper translated the original Japanese into the vernacular while leaving some familiar Chinese characters in their original form. The fact that Humane Elixir was specifically manufactured under the direction of two Japanese "Ph.D.s" (*paksa*) was highlighted, alluding to a novel authority. At the same time, familiar terms of traditional medicine, such as "elixir" (*tan*) and "original vitality" (*wŏn'gi*), evoked the body in tension between fulfilled repletion and sickly depletion. The advertisement was visually concise and pleasantly presented, displaying eclectic features of production through which multiple linguistic resources and clinical reasoning intersected.

Absent rigid regulation, advertisements for Humane Elixir boasted various designs and made frequent appearances.[4] A holiday mood, like that of New Year's Day, was used for special promotion. Children's health was supposed to be improved with Humane Elixir. The fact that six thousand doses of Humane Elixir were sent to China in 1914 to aid flood survivors was highlighted. When the Competitive Exhibition (Kyŏngsŏng kongjinhoe) was held in Seoul in 1915, the signboard for Humane Elixir was put on the clock tower, representing its leading position—and this event was used for newspaper ads. Its popularity abroad was also highlighted: Humane Elixir appeared in a novel that proved popular in India, and four overseas sales branches were touted as evidence of its international fame. When Li Yuanhong (1864–1928) became the president of the Republic of China in 1916, the portraits of him and Vice President Feng Guozhang (1859–1919) were featured with an endorsement for Humane Elixir in their own handwriting.[5] Without exaggeration, Humane Elixir dominated the Korean newspaper's advertising space during the first half of the twentieth century in terms of sponsorship and publicizing techniques.

Japanese medical commodities in general were competitively advertised in Korean newspapers between the 1910s and 1930s. Medicine-related ads occupied 30 percent of advertising space in *Tonga ilbo* (East Asian Daily) in 1923 and accounted for more than half by 1938. The volume of Japanese ads in the same newspaper increased from 36.1 percent in 1923 to 61.6 percent in 1938. The rising proportion of Japanese ads, among which those for medicine were most prominent, is also found in the *Daily News* and *Chosŏn ilbo* (Korean Daily).[6] The prominence of medical ads partly reflects the vibrant sales force of the pharmaceutical companies and major apothecaries. Japanese manufacturers, who successfully navigated the Meiji government's effort to control patent medicines since 1870, aimed to fully explore commercial opportunities in Korea after the country signed an unequal treaty with Japan in 1876. Japanese businessmen such as Arai Kotarō (b. 1867) and Yamagishi Yūtarō (1868–1927) began to establish their apothecaries in Korean cities after the Russo-Japanese War (1905) and were followed by wholesalers and retailers.[7] By 1906, fifteen Japanese-owned pharmacies ran their businesses in Keijō (Seoul) alone, and the value of Korean imports of patent medicine from Japan increased from

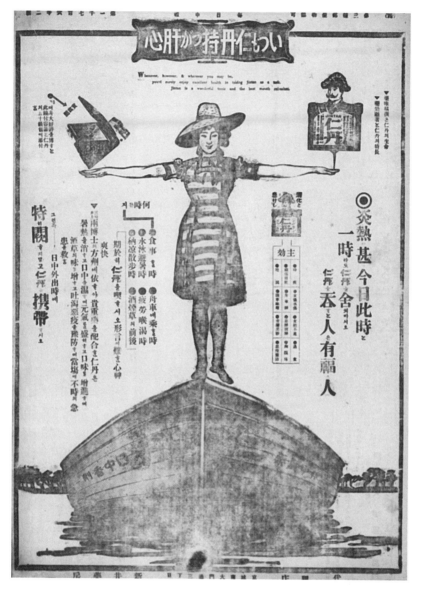

FIGURE 1. An advertisement for Indan, Humane Elixir (*Maeil sinbo*, September 1, 1911, p. 4)

¥105,515 to ¥235,738 between 1911 and 1917.[8] Japanese manufacturers (and their ads) successfully made their way to Korea beginning in the late nineteenth century, grew between the Russo-Japanese War and Japanese annexation (1910), and reached their heyday in the 1930s.

How, then, did Korean manufacturers respond to Japanese dominance in the marketplace? Did Korean drug sellers and pharmacists try to emulate the Japanese initiative, or did they plot their own marketing strategies? Beginning with these questions, this chapter focuses on the changing measures of Korean manufacturers in using indigenous features for fashioning medical advertisements. To what extent did Korean manufacturers formulate particular traits of Korea in securing an advantage for their business? In what ways did sensitivity to native culture parallel the cosmopolitan attributes associated with consuming medicine?

Advertisements as a Lens

To trace Korean manufacturers' efforts at legitimating local products, this chapter primarily analyzes medical advertisements in vernacular newspapers.[9] In 1886, the first advertisement by a German trading company, Sech'ang yanghaeng, emerged in the weekly *Hansŏng chubo*, which was rendered in Korean with classical Chinese mixed in.[10] After 1896, medicine-related commodities such as quinine were sporadically publicized in Korea's first private and vernacular newspaper, *Tongnip sinmun*, and its English edition, the *Independence*, along with tobacco, Western fabrics, dyes, and books on English grammar. The budding space for medical advertisements in a couple of Korean newspapers, however, mostly vanished after the Japanese annexation in 1910. Only the pro-Japanese paper *Daily News* survived, the successor to *Taehan maeil sinbo* (Korean Daily News), and continued to be circulated until 1945. In the 1920s, a wave of nationalist journalism saw the establishment of the *Korean Daily* and the *East Asian Daily*, but after a couple of suspensions by the colonial surveillance, both were forced to close in 1940.[11] Between 1929 and 1939, the circulated copies of the *Korean Daily* and the *East Asian Daily* increased from 61,288 to 115,371.[12] Certainly, outdoor billboards and dozens of newly launched journals provided a realm for publicizing commodities, yet in terms of circulation and recognition, the three vernacular newspapers provided the most stable space for engagement if an advertiser anticipated Korean responses.[13]

Obviously, fully tracing consumers' choices through advertisements by themselves is difficult. We cannot demonstrate the impact of flamboyant illustrations on consumers, nor can we assess the credibility of information deployed by merchants for commercial causes. However, as Douglas E. Haynes persuasively argues, advertisers at least displayed certain kinds of knowledge and values to appeal to their imagined audiences. Advertisers' sensitive "antennae" often detected murmuring anxieties and emerging desires, then made efforts to "shape new needs and reorient old ones."[14] Advertisements in historical context did not just expose false consciousness in operation; they reflected the producers' technique of (re)directing the existing knowledge and meanings.

According to Sherman Cochran, considering the agency of advertisers and manufacturers of medicine helps us fathom the complicated procedure of localization from nonelites' point of view. Outside the conventionally perceived intellectual realm, Chinese entrepreneurs, apothecary owners, and talented traders actively appropriated Western thoughts and values. Compared to their elite counterparts, the merchants and manufacturers exhibited a less cosmopolitan mode of thinking, but they "played prominent roles in the localization of Western ideas and images in China." This process of indigenization from the bottom up challenges an ahistorical bifurcation of "Chinese" and "Western" medicine and makes us aware of the rich and complex composition of "Sino-Western or traditional-modern dualism" in everyday life.[15]

The analysis of Korean advertisements takes advantage of Cochran's thematic approach to localization. The Korean maneuvering of multiple resources in the early twentieth century enriched the meaning of the "local" at the juncture of unprecedented possibilities and constraints. Korean agents translated Japanese originals into Korean, then gradually replaced the Japanese images with Korean faces and costumes to a certain extent. Ideas about Korean lifestyles, familiar flavors, major political events, and common complaints of bodily discomforts were selectively detected and then used for efficient advertising. Overall, eclecticism governed the graphic and verbal composition, yet some companies yearned to deliver didactic messages. Closer scrutiny of Korean advertisements can reveal nuances and tone in the existing narratives of localization.

Korean apothecary owners and dealers, unlike their Chinese counterparts, ran their business under the formal control of Japanese colonialism. Did this indicate harsher surveillance or more commercial opportunities? In her analysis of the construction of consumer identity among Koreans in 1930s Seoul, Se-mi Oh highlights the dual conditions of creating con-sumer culture. Beyond an oppressive grip, the Japanese empire ultimately controlled Korean urbanites by producing the "desiring subjects through the visual display of modernity."[16] The newly established urban space, the bustling Japanese commercial district, and luxury items invited Korean urban residents to conform to cosmopolitan consumerism without a sub-stantive promise for political emancipation. The Japanese creation of desir-ing subjects, according to Oh, was inclusive, anticipating the participation of Koreans as "cultural producers in their own right."[17] Although Oh's analysis focuses on the newly emerging urban space of Seoul and only includes "luxurious items," the performative nature of emerging consumer-ism in the 1920s and 1930s helps us read the Korean manufacturers' agency outside the duality of pro- or anti-Japanese sentiment.

By examining early twentieth-century advertisements of medicine, this chapter primarily complicates Korean manufacturers' efforts at local-ization on their own terms. The bottom-up construction of the "indigenous and foreign" dualism complements the three previous chapters, which mostly focus on elites' expressions and practices. Inasmuch as educated physicians and scholars in the Chosŏn dynasty longed to recognize the names of local botanicals, which would thereby project their position in the world, merchants of medicine enthusiastically (re)titled hundreds of medical commodities, anticipating new connections and circulation. As the articulation of Eastern medicine displayed a will for cultural distinc-tion, a successful Korean entrepreneur envisioned the modern identity of Koreans through the production and consumption of biomedical com-modities. As the use of the "Chosŏn" label in early twentieth-century scientific discourses complicated the interaction between Japanese and Korean doctors of biomedicine, so did the medical commodities expose the multilayered interactions between Japanese and Korean merchants. Before fleshing out these points, Japan's domination in marketing medi-cine is explained more fully.

Japan's Domination of the Korean Market

Japan had a significant impact on the restructuring of the pharmaceutical market in Korea, primarily by enhancing legal authorization. In March 1912, the Japanese Government-General in Korea (Chōsen sōtokufu) promulgated the Regulations for the Enforcement of Laws for Medicinal Products and the Medicinal Product Businesses (Yakuhin oyobi yakuhin eigyō torishimari shikō kisoku). This law was intended to specify four categories of pharmaceutical professions, identifying their quality of knowledge and degree of commercial engagement: the pharmacist (*cheyaksa*) was supposed to "combine chemicals according to doctors' prescriptions"; the manufacturer (*cheyakcha*) to "prepare and sell medicinal products"; the drug seller (*yakchongsang*) to "sell chemicals"; and finally, the merchant of patent medicines (*maeyagŏpcha*) to "sell patent medicines by manufacturing, introducing, or importing."[18] Before the Japanese law, the Great Han Empire (Taehan cheguk, 1897–1910) in January 1900 also attempted to distinguish medicinal preparation from its sales, thereby controlling the murky overlap of diagnoses, prescriptions, and trades.[19] To the Japanese colonial government, however, the Korean regulations had largely been nominal, without any legal ramifications.

The bottom line of the newly enhanced colonial law aimed to distinguish biomedically trained pharmacists from practitioners of traditional medicine and herb dealers, as well as from those who merely handled chemicals and poisonous substances. In reality, boundaries continued to be blurred even between pharmacists and doctors. Diagnoses and consultations frequently occurred at apothecaries, whereas doctors and their clinics continued to be involved in medicinal trades in some way or another. Herb dealers had easy access to Western medicinal items. Due to these overlaps, the colonial government and its police force sought to enforce its regulations, and reported violations could incur the cancellation of business permissions, licenses, and patent registrations. Despite sporadic complaints and criticism from Korean practitioners of traditional medicine, the newly imposed regulation remained intact until 1945.[20] Korean manufacturers and dealers of pharmaceuticals needed to (re)structure their knowledge, practices, and identity to maintain their businesses and protect their careers under the colonial government.

To enhance the effectiveness of its legal restrictions, Japan facilitated the flow of biomedically trained people and medical commodities by establishing modern hospitals in treaty ports in Korea. Clinics run by Japanese army doctors were first established in Pusan in 1877, Wŏnsan in 1880, and Seoul and Inch'ŏn in 1883. Called Government-Sponsored Hospitals (Kwallip pyŏngwŏn), these institutions aimed to "relieve [Japanese] settlers' disease and to provide hygiene."[21] Compared to the Government-Sponsored Hospitals, the Seoul Hospital (Hansŏng pyŏngwŏn), established by Japan in 1895, more explicitly displayed imperial benevolence toward Koreans in need. Most Korean patients were treated for free or at low cost. On annexation in 1910, the Government-General Hospital replaced the Hospital of Great Han (Taehan ŭiwŏn), which was established by the Japanese Residency-General in Korea (Chosŏn t'onggambu) in 1907. Paralleling the initiative in Seoul, branches of Charity Hospitals (Chahye pyŏngwŏn) began to be built nationwide in 1909; these regional hospitals were renamed Provincial Hospitals (Torip pyŏngwŏn) after 1925. Overall, forty-one provincial hospitals had been established in Korea by 1938.[22] Through these facilities and their networks, biomedically trained Japanese pharmacists were introduced. Approximately fifty Japanese pharmacists worked in Korea around 1910. After a period of service, they frequently opened apothecaries and made a variety of Japanese medicinal commodities available for purchase to the public, including cosmetics and miscellaneous goods.[23]

Yamagishi Yūtarō's Tenyūdō, one of most famous Japanese-run apothecaries in Seoul, first established its wholesale shop in 1906 and prospered by supplying medicine and medical equipment to the Charity Hospitals. In 1915, the Yamagishi apothecary employed a staff of thirty and had sales of ¥500,000 a month; sales rose to ¥1 million a month in 1935.[24] As the Japanese imperial vision for Manchuria took shape with the establishment of Manchukuo in 1932, Korean branches were regarded as a significant bridge to the Chinese continental market. Based on the booming business during the 1920s and 1930s, Japanese apothecaries and pharmaceutical companies began to construct factories in Korea to produce agricultural chemicals, tonic soups, and medicines that used Korean raw materials and special products.[25]

The Japanese initiative in Korea was supported in a variety of ways, primarily by using various associations of Japanese pharmacists and drug merchants to consolidate the Japanese network in Korea. The Association of Pharmacists in Korea (Han'guk yakchesahoe) was first established in 1909 under the leadership of Japanese pharmacists who worked for the affiliate hospital of the residency-general. This organization's membership was limited to those holding a license in biomedical training. Among twenty-eight regular members, only one was Korean. The association was renamed the Pharmaceutical Society in Korea (Chosŏn yakhakhoe) in 1913 in an attempt to recruit anyone associated with the pharmaceutical field, and it held lectures to publicize newly established qualifications for pharmacists. The association provided information for pharmacists about the regulation of toxic chemicals and specified principles for preparing biomedicine. A short course managed by the society was developed into a training school, which later became the School of Pharmacy in Korea (Chosŏn yakhakkyo).[26]

Some major newsletters fortified the Japanese-centric network: *Keijō yakuhō* (Newsletter of pharmacy in Seoul) was first published in 1915 and sponsored by the Arai and Yamagishi apothecaries. As trade prospered, Arai also funded *Mansen no keshōhin shōhō* (Commercial news of cosmetics in Manchuria and Korea) and the *Medical World of Manchuria and Korea*. Stimulated by the good response to these ventures and the successful impact of the *Newsletter of Pharmacy in Seoul*, drug manufacturers in Pusan published *Chōsen yakuhō* (Newsletter of pharmacy in Korea) in 1927.[27]

Given the aforementioned Japanese initiative, it is not surprising that Korean salesmen struggled to secure sponsorship from major Japanese pharmaceutical companies.[28] Those who were once affiliated with the Sales Department of major Korean newspapers testified to how desperately they solicited Japanese support. According to Cho Yong-man (1909–95), who worked at the *Daily News* in the 1930s, "We Korean newspapers relied heavily on the sponsorship of Japanese pharmaceutical companies for a while."[29] Kim Sŭng-mun (1901–83), who was hired at the *East Asian Daily* in 1928 and worked there for thirty-eight years, described how he urgently sought Japanese business relationships by knocking on every door at an advertising agency. To guarantee substantial support, he stayed in Tokyo for more than six months, personally hired a local Korean who knew the Japanese advertising business, and then struggled to establish personal relationships with Japanese advertisers.[30] In another case, Sŏ Hyang-sŏk (1900–85), who also worked at the *East Asian Daily* during the 1930s, confessed his mixed feelings regarding the overwhelming number of Japanese ads in nationalist Korean papers. A print medium aimed at hailing Korean nationalism became the major means of introducing Japanese commodities.[31]

In summary, Japan played a significant role in reshaping the early twentieth-century medical market in Korea by regulating the modern license system, introducing biomedicine-centered institutions, establishing a network of personnel, and providing a venue for publicity. What, then, were Korean merchants' strategies for survival and success? We can trace a pattern of success and setbacks in the creation of a Korean sales force through the patent medicines' newly named water, elixir, pill, and paste.

Patent Medicines and Strategies of Localization

WATER, HWALMYŎNGSU

Lifesaving Water (Hwalmyŏngsu), a liquid digestive, was first manufactured in 1897 by Min Pyŏng-ho (1858–1939), who was from a *yangban* family and held an official position after passing the state examination.[32] As a converted Christian, Min Pyŏng-ho came to nurture his interest in Western medicine while maintaining his prowess in scholarly medical tradition. He first extracted a few already well-known herbs, such as *Atractylis koreana nakai* (or Korean atractylodes, *ch'angch'ul*), dried orange peel, and silver magnolia, then added imported catechu and dried clove camphor. Newly introduced supplies like alcohol and menthol were added, creating novelty as a foil for more established concoctions.[33] This resulted in Lifesaving Water, which was supposed to be effective for "urgent catarrh without distinction of sex or age, children's vomiting and diarrhea, epidemic diseases, acute indigestion, and chronic diarrhea."[34] Although indigestion was primarily targeted, a broader range of symptoms, like catarrh and even cholera-like epidemics (*hoyŏksŏng chirhwan*), was used to solicit sales. In 1910, the list price for a sixty-milliliter bottle was forty *chŏn*, approximately equal to sixteen dollars today.[35] Compared to the item's contemporary price—eighty cents per seventy-five-milliliter bottle in South Korea—the original price was quite expensive. Despite the high cost, the news of the water's "miraculous" efficacy quickly spread outside Min Pyŏng-ho's residential area. His Tonghwa Apothecary (Tonghwa yakpang) prospered, and he registered ninety-eight patent medicines by 1908.[36] Although Min Pyŏng-ho was neither trained in biomedicine nor from a merchant family, he successfully exploited the market by combining his understanding of traditional medicine with knowledge about Western medicine.

Min Pyŏng-ho exhibited both strength and weakness as an advertiser. A fan-shaped trademark registered in 1910 turned out to be successful. The brand image distinguished Lifesaving Water from dozens of knockoffs promptly emulating its success. Min Pyŏng-ho was known to say the fan-shaped image was inspired by a verse from *Sigyŏng* (Book of odes).[37] Inasmuch as the harmony of bamboo and paper create the clear wind of vital *ki*, Min Pyŏng-ho hoped his synthesis would produce the unprecedented virtue of remedy. Inasmuch as the Humane Elixir's mustachioed man effectively symbolized the brand, the fan-shaped figure created a successful visual representation of the apothecary's distinction, thereby contributing to the publicity of Lifesaving Water. In most printed ads, however, the Tonghwa Apothecary fell short of fully exploiting the graphic element of advertisement. Most of its ads in the *Daily News* during the 1910s failed to use the printed space for visual attraction, focusing only on textual information.[38]

The advertisements represented the apothecary's strong nationalism. The founder's son, Min Kang (1883–1931), expanded and developed his father's apothecary starting in the 1910s and was passionately involved in the Korean independence movement. His anti-Japanese activism eventually put his apothecary in danger, as sixty-three out of eighty-seven medicine patents were canceled by the Japanese colonial government in the early 1920s.[39] Tonghwa's ideological inclination did not perish after Min Kang passed away. Celebrating the Korean marathon runner Son Ki-jŏng's (1912–2002) gold medal at the Berlin Olympics in 1936, the apothecary ran a special advertisement. With the startling caption "Sons of the (Korean) Peninsula, Soaring Vigor," the ad highlighted the significance of a healthy stomach as the foundation of a healthy Chosŏn Korea (Figure 2).[40]

Under Japanese colonial governance, the apothecary's explicit expression of nationalist consciousness was an exception rather than a convention. The majority of Korean manufacturers of patent medicines showed a cooperative attitude toward the Japanese-led hygienic modernity and attempted to franchise the Japanese commercial initiative.[41] A series of ads put in the pro-Japanese newspaper *Daily News*, such as one by Yi Kyŏng-bong's (n.d.) Chesaengdang Apothecary (Chesaengdang yakpang), publicly celebrated the Japanese annexation of Korea in 1910. The apothecaries' ads generally promoted the idea of hygiene, thereby urging Koreans to wholeheartedly embrace the Japan-led civilization and enlightenment.[42]

The advertisement of Lifesaving Water in the 1910s showcased the persistent elements of traditional medicine in the synthesis of "novel" medical commodities. Exposure to Western medicine and modern advertising techniques surely helped the Min family launch a new brand. Yet

FIGURE 2. An advertisement for Hwalmyŏngsu, Lifesaving Water (*Chosŏn ilbo*, August 11, 1936, p. 4)

with sensitivity to market change and policy promulgations, Min Pyŏng-ho deployed the strategy he had been successful with by that time: making globules and waters out of familiar herbs, responding swiftly to his neighbors' favorite flavors, and publicizing their brand of manufactured medicines according to market changes. Under the colonial government's explicit intention to authorize only biomedicine, Min Pyŏng-ho's success demonstrates a skillful maneuvering of traditional medicine in expanding his apothecary's inventory of products developed through the synthesis of the traditional and the modern.

ELIXIR, CH'ŎNGSIM POMYŎNGDAN

Another example of success was the Elixir for Clearing the Heart and Guarding Life (Ch'ŏngsim pomyŏngdan), manufactured by Yi Kyŏng-bong. He had engaged in the herb trade with China for years, and was inspired by the success of the Japanese Humane Elixir. Instead of importing the Japanese items, Yi Kyŏng-bong determined to manufacture his own brand. His elixir was made from herbs like borneol and peppermint, tasted sweet, and consisted of small red granules. As a domestic product, the elixir was sold for five *chŏn*,[43] half the price of the Humane Elixir. Yi Kyŏng-bong's handsome, Westernized appearance, combined with his eloquent and flamboyant style of speaking, added a modern image to this medicine. Frequent street performances with music and colorful banners promoted the medicine and the apothecary, providing a kind of public entertainment. Yi Kyŏng-bong's Chesaengdang Apothecary was first established in one of the treaty ports, Chemulp'o, then moved to Seoul.[44]

The printed ads for Yi Kyŏng-bong's elixir showcased the manufacturer's ability to pick up the latest advertising techniques and the conventional mode of expressing health concerns. As the main product of the apothecary, the elixir was aggressively publicized. Between 1907 and 1910, for instance, the *Korean Daily News* advertised the elixir more than five hundred times.[45] Given that the paper was published twenty-four or twenty-five days a month, the apothecary planned an almost daily appeal to readers. The ad was not visually attractive in its first debut in 1907. Following the convention of the time, it merely presented the business title, brand name, and major efficacy. In the following versions, the ad trimmed wordy information, showcased an image at the center, and presented the selling points in a brief poem in a style familiar to most Koreans.[46] Compared to others at that time, the elixir's ad layout was quite extraordinary. The apothecary understood how to use the brand name in Chinese characters as a graphic, not merely as a signifier (Figure 3). In addition, one of the ads in 1911 used an image of an airship with a gentleman of modern appearance, signaling the apothecary's ambition to reach out internationally (Figure 4). The copy read, "An airship arises from the ground, conquering the vast skies of *ŏk* [one hundred million], *ch'ŏn* [one thousand], *man*

FIGURE 3. An advertisement for Ch'ŏngsim pomyŏngdan, the Elixir for Clearing
the Heart and Guarding Life (*Taehan maeil sinbo* with Chinese and Korean
combined, November 14, 1909, p. 3)

[ten thousand] -*ri*, the Elixir for Clearing the Heart and Guarding Life
explores abroad, conquering the *ŏk, ch'ŏn, man* people's stomachs."[47]

In addition to the refined technique of visualization, words in the ad
signaled the manufacturer's familiarity with the expressions of illness
among fellow Koreans. *T'ojil*, in the second stanza of the poem, presented
the place-specific understanding of sickness. It read, "Every winter, many
suffer from phlegm and cough, which is *t'ojil* of our country. Not to mention other years, last winter, those who experienced the elixir's efficacy
reached tens of thousands."[48] *T'ojil* literally implies a disease caused by the
inappropriate relationship between water and soil. In the advertisement,
which highlighted "our country" and "last winter," *t'ojil* connoted a time
and place specific to the local imagination of illness. Interestingly, *t'ojil* in
the 1920s and 1930s was used to indicate pulmonary distomatosis, a rampant parasite infection in the lungs. The *Daily News* frequently reported
the Japanese colonial government's efforts to control *t'ojil*. It is unknown
why *t'ojil* became the native term for pulmonary distomatosis, yet its endemic prevalence likely borrowed the place-specific connotation of
t'ojil.[49] What seems intriguing here is not the exact definition of *t'ojil* as
a clinical entity but the advertiser's sensitivity to the popular understanding of an illness rooted in the local environment.

Following the mention of *t'ojil*, in the next stanza, the idea of place-
specific etiology was expressed more literally with the term *sut'o pulbok*,
a state of disequilibrium between water and soil.[50] The identity of this
illness category is also elusive, yet the term *sut'o pulbok*, which indicated

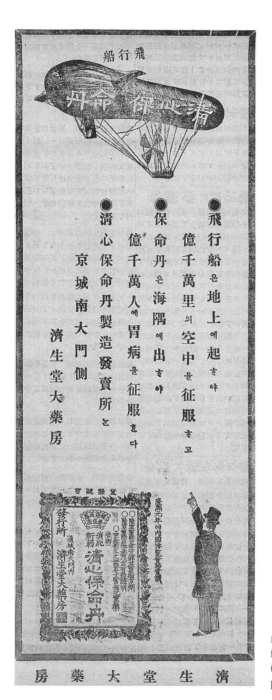

●淸心保命丹製造發賣所ᄂᆫ
京城南大門側
濟生堂大藥房

●保命丹은海隅에出ᄒ야
億千萬人에胃病을征服ᄒᆫ다

●飛行船은地上에起ᄒ야
億千萬里의空中을征服ᄒ고

FIGURE 4. An advertisement for Ch'ŏngsim pomyŏngdan (*Maeil sinbo*, April 28, 1911, p. 4)

disharmony in local conditions, was rendered many times in the ads in parallel with other illness categories such as indigestion, (minor) intoxication, bad breath, summer heat and winter cold, and motion sickness. The "dissatisfaction of water and soil" posited a central selling point of the Elixir for Clearing the Heart and Guarding Life, and the term appeared in almost every published ad for the product between 1907 and 1910.

The advertisements of Yi Kyŏng-bong's apothecary hint that his sales force shared his recognition of the commercial potential of Japanese patent medicines and Korean aspirations for novel visual culture. Simultaneously, his use of *t'ojil* and *sut'o pulbok* reveals the popular imagination of illnesses, which was caused by the locally specific interaction among soil, water, and people. Whereas Min Pyŏng-ho's Lifesaving Water encouraged the use of an urgent countermeasure for mostly acute problems of indigestion, Yi Kyŏng-bong's elixir reminded Korean consumers of the need to constantly manage bodily discomforts, which might be caused by the changing configuration of the local environment.

PILL, PAEKPOHWAN

Medical advertisements aimed at Koreans were circulated in 1930s Manchuria as marketing networks expanded.[51] The design and visual register of the ads, published in either *Mansŏn ilbo* (Daily News of Manchuria and Korea) or *Medical World of Manchuria and Korea*,[52] were not entirely dissimilar to their domestic counterparts. Yet a series of ads demonstrated their sensitivity to the Manchurian environment. Among the many medical commodities advertised, one of the most noticeable was general replenishing pills, such as the Pill of Nourishment with One Hundred Ingredients (Paekpohwan), made by the P'yŏnghwadang corporation, or the Longevity Pill of Nourishment with One Hundred Ingredients (Mansu paekpohwan), made by the Asia Great Apothecary (Asea taeyakpang).[53] These replenishing pills surely were advertised in domestic newspapers, yet in Manchuria, the size and frequency of the ads were expanded, partly due to lower advertising costs. A full-page ad with a bonus gift, such as a comparison table of solar and lunar calendars, was easily found in the *Daily News of Manchuria and Korea* (Figure 5).[54]

Geography mattered in imagining the efficacy of the consumed medicine. The Longevity Pill of Nourishment with One Hundred Ingredients, for instance, highlighted the fact that traditional medicine had originated in Manchuria, then a part of China, and not Korea or Japan. This alludes to

FIGURE 5. An advertisement for Paekpohwan, the Pill of Nourishment with One Hundred Ingredients (*Mansŏn ilbo*, January 3, 1940, p. 3)

an image of a secret formula, which might guarantee a greater excellence than those newer products from Korea or Japan. Although the apothecary manufacturing the pill did not exclusively belong to Manchuria, this store emphasized geographical difference to highlight the Chinese origin of traditional medicine and thus appeal to Koreans in Manchuria.[55] The ad also reminded readers that the recently promulgated regulation limited the export of herbs from Japan and Korea, which possibly legitimated the consumption of made-in-Manchuria medicines. A few days later, an article in the newspaper reported the enhanced wartime control of the medicinal trade and the reactions of herb dealers in Korea.[56]

Furthermore, the advertisements frequently recalled the harsh environmental conditions peculiar to Manchuria, such as drought and extreme temperatures that sapped water, energy, and primal vitality from plants and people, to point to the origins of a perceived decline in immunity suffered by Koreans in Manchuria. Even those who had never taken the replenishing medicine should consider taking the pill soon.[57]

A corollary of the portrayal of Manchuria as a place of rampant illness was to depict Korea as the best location for herbs. The Pill of Nourishment with One Hundred Ingredients was apparently the best in the world because its ingredients were rooted in Korean soil. Unlike the medicines imported from the West or other places in East Asia, this pill was made from mysterious grass roots from the homeland and the fruit of trees free from all types of chemical compounds and foreign contamination. Korean soil was an indispensable element of health, even for those who resided elsewhere. The ad mentioned that the Imperial University, under the support of the government-general, launched research into Korean indigenous medicine, examining botanicals available in Korean soil.[58] Whereas a certain brand described how medicine from Manchuria (China) guaranteed better efficacy to attract Korean immigrants in Manchuria, others emphasized that Koreans needed medicine whose ingredients were exclusively grown in Korea.

More important, the advertisements' mode of describing health problems often paralleled journalistic comments or literary expressions about sickness among Korean immigrants. Politically marginalized and culturally isolated, Koreans in 1930s Manchuria often complained about the lack of medical infrastructure and accumulated psychosomatic discomforts, which hardly found immediate solutions. Korean immigrants were viewed as vulnerable to seasonal ruptures, chronic pains, and emotional instability. A series of articles called a "Survey of the Frontier Towns of Korean Settlers in Northern and Southern Manchuria" mentioned the urgent demand for hospital construction, which implied a lack of available medical facilities.[59] The "Urgent Problems of Our Region" stressed the prevalent illness caused by the unfamiliar "climate and the relation of water and soil" in Manchuria.[60] A series of articles on "Life of Koreans in Manchuria Viewed from Women's Perspective" interviewed some Korean women and detailed their bodily discomfort caused by the physical and cultural environment of Manchuria.[61] For these women, the loneliness in a foreign land, accompanied by ethnic contestations, exacerbated the degree of a depressive mode. The family medicine section also reported the frequent incidence of fever in children, of which the cause was not straightforward. The children's fevers, according to the article, often developed into acute pneumonia.[62] Overall, the journalistic gaze captured the mundane experience of severe cold, muscular fatigue, and mental exhaustion, whose symptoms were neither curable by any prompt countermeasure nor treated in adequate institutional settings.[63]

If the articles in the pro-Japanese newspaper described the immigrants' health issues mildly, the well-known Korean woman novelist Kang Kyŏng-ae (1906–44) vividly depicted the socioeconomic root of the immigrant women's psychosomatic exhaustion in "Sogŭm" (Salt).[64] At the intersection of class, gender, and ethnic contestation, the protagonist loses her entire family in a series of sudden, violent incidents. Although Kang Kyŏng-ae did not delve into illness as an analogy, the embodied loss and despair of an impoverished Korean woman fully reminds us of the hardship through which the imagined consumers of Manchuria might manage their daily lives in this time period.

The frequent appearance of replenishing medicines, or tonics, indicates that the market space existed in the vacuum of the public health infrastructure for Korean immigrants in Manchuria. Targeting both chronic and acute diseases, promising prevention and cures, and encompassing indigenous and biomedical terminologies, the ads came to terms with the Korean experience of illness, which had otherwise met with no adequate solutions. If read synchronically with newspaper articles, individual testimonials, and literary compositions of 1930s Manchuria, the medical advertisements targeted at Korean immigrants reveal more than the concepts of desire, pleasure, and fashion, urging us to conjecture about the depth of undocumented illness experience from the bottom.

PASTE, CHOGOYAK

Successful manufacturers and herb dealers relied heavily on local brokers. The marketing techniques of the Ch'ŏnil Apothecary (Ch'ŏnil yakpang) exemplified the increasing sales promotion in the 1920s and 1930s, which included premiums and spectacles for the middleman. The Ch'ŏnil Apothecary was first established in 1913 by Cho Kŭn-ch'ang (active 1910s–30s), who specialized in tumors.[65] Given the lack of modern surgical treatment, most abscesses were handled by lancing. A range of pastes, such as the one manufactured by the Cho family, were applied to the incised part. The Cho medical paste gained a reputation in Seoul at the end of the nineteenth century, yet this story of a successful origin might be ascribed to the apothecary's commercial achievement in the early twentieth century. Cho's son, who received a modern education in Japan and launched a new herb trade there, commercialized the family plaster by using the latest sales strategies.

The Ch'ŏnil Apothecary aimed at solidifying domestic sales networks. It published a monthly newsletter, *Ch'ŏnil yakpo* (Medical newsletter of Ch'ŏnil), which circulated among traders nationwide. Special events for middlemen or local sales people were held frequently, and free gifts were often distributed. Calendars printed in Japan and a variety of premium items, such as playing cards or the four-stick game, attracted local traders. It was not unusual to invite local traders to one of the most flamboyant restaurants in Seoul, which usually provided scores of singing girls. Famous singers, a comic storyteller, and a dancing master put on performances to please local dealers.[66] The highlight of the apothecary's special sales events in Seoul, riding an airplane, lasted only five minutes, yet more than eighty traders gathered. A trip to northeastern China proved popular, too. Eight days in duration, the trip included major cities in northern and southern Manchuria as well as Beijing and Tianjin. The apothecary paid all travel and accommodation expenses. This type of special sales promotion continued until 1939.[67]

The costly promotions eventually paid off. The low cost of manufacturing and its profitable margin induced the influx of undifferentiated brands of patent medicines into the market. Newspapers criticized their low quality and the excessive competition among apothecaries. Whereas the retail shop usually took a 40–50 percent profit margin, latecomers who attempted to market knockoffs had to provide bargain prices, allowing up to 60–70 percent markup.[68] The initial cost for manufacturing the medical plaster of the Cho family was less than two *chŏn*, which then rose to ten *chŏn* retail and five *chŏn* wholesale.[69] Given that over 1,500 kinds of patent medicines were being sold in Korea by the 1920s, what distinguished one from another was not its "scientifically" proven medical efficacy, but the degree of publicity combined with exciting gifts, special events, and an unprecedented scale of spectacles. The Ch'ŏnil Apothecary well understood the promise of local men, so it did not spare any expenses in creating big events for traders.

Min Pyŏng-ho's Tonghwa Apothecary also made a special effort to administer to brokers. Min established a hierarchical branch network according to its sales volume within the first ten years of opening. Tonghwa's publicized regulations of sales in 1911 identified three levels of offices: regional, district, and provincial. In 1910, the apothecary worked with 108 regional and district branches in one province, P'yŏngan. In addition, special sales offices, which monopolized the apothecary's items as wholesalers, were identified. In 1919, the apothecary celebrated 186 special sales offices nationwide.[70]

With the effort to enhance the sales network, patent medicines were handled by a variety of traders. A newspaper article in 1923 counted 350 manufacturers of patent medicines, 150 importers of Japanese medicines, and 2,543 wholesalers of medical commodities in just one province, Kyŏnggi-do, the surrounding area of Seoul. The article estimated the total sales amount of patent medicines to be ¥800,000. As the total of yearly medical consumption in the province was estimated to be ¥5,000,000,[71] clearly the sales of Western medical commodities and herbal raw materials outnumbered that of patent medicines. However, the quantified sales scale demonstrates the expanded market share of the patent medicines up to the early 1920s.

Min Pyŏng-ho, Yi Kyŏng-bong, and other successful drug sellers, whose products bore the novel names of "water," "elixir," "paste," and "pill," shared the following characteristics. All were versed in herbs and became experienced traders in one or two generations. Patent medicines from China and Japan inspired herb traders to produce their own brands by adding particular flavors, appearances, and promotions. With the exception of Min Pyŏng-ho, whose major market was initially in P'yŏngyang, the apothecaries first began in treaty ports and then moved to a main street in the center of Seoul. Hardly any biomedical principles or ingredients were involved, yet by finishing their products with a modern gloss, herb traders envisioned a novel synthesis. The advertisements of the successful patent medicines testified to their sensitivity to familiar flavors, patterns of illness, local trade networks, and prevalent health concerns among Koreans, showcasing varied forms of localization in their own terms.

Combining Nationalism with Biomedicine

Manufacturing biomedicines gave Koreans another opportunity to establish their own commercial networks. The Yuhan Company (Yuhan yanghaeng) successfully launched a pharmaceutical company that facilitated nationalist sentiment and ideals of enlightenment through its advertisements.[72] Under the leadership of its founder, Yu Ir-han (1895–1971), the Yuhan Company aggressively publicized its products and explored a new path for combining commodities with ethical messages. The company's didactic framing of health management, consumption, and industrialization contrasted and overlapped with the patent medicines' use of indigenous attributes in advertising.

YU IR-HAN'S PATH TO
THE YUHAN COMPANY

Yu Ir-han was born in 1895 as the first son of Yu Ki-yŏn (n.d.), a promising merchant and converted Christian in P'yŏngyang. Impressed by Western missionaries' medical services and educational efforts, Yu Ki-yŏn decided to train his son abroad to avoid the impending rule of the Japanese empire. At the age of nine, Yu Ir-han was sent to San Francisco, then to high school in Nebraska. After majoring in business at the University of Michigan, Ann Arbor, Yu Ir-han briefly worked at General Electric. He soon envisioned his own business while gradually tapping into the emerging US demand for "Oriental" flavors, commodities, and food ingredients. After Yu Ir-han succeeded in producing canned bean sprouts, which were indispensable for Chinese cuisine yet difficult to preserve in a raw form, his business surged. Yu Ir-han established La Choy Food Products in the United States in 1922 with his friend Wally Smith (n.d.). Within four years, the company was grossing $500,000 annually and expanding its commercial network throughout the major US cities.[73]

To secure good-quality mung beans, Yu Ir-han traveled to China and Korea in 1925. Twenty-one years had passed since he left Korea as a boy. Passing through southern China, Korea, and Manchuria, he came to better understand the socioeconomic circumstances of his motherland. According to most biographies, this strenuous trip to East Asia motivated Yu Ir-han to return to Korea permanently. The "tragic backwardness of Korean medical conditions" looked more serious than the young businessman had imagined. The lack of appropriate medicines for curing "seasonal epidemics, quacks and all kinds of parasites, tuberculosis, malaria, and skin diseases" touched his emotions, and his "passionate love for his own country" urged him to give up his promising career in the United States and return to Korea.[74]

The image of a poor and weak Korea waiting for a Korean-born US businessman's benevolence, however, requires a balanced evaluation. After over a million Koreans joined in the March First Movement calling for the country's independence in 1919, the Japanese Government-General in Korea put forth the cultural rule, hiding its oppressive grip under the desk. The economic implication of this changed policy was the complete abandonment of the Company Law in 1920. Earlier, in 1910, the government-general had aimed to strictly control both Korean and Japanese private investment in Korean industry by promulgating the Company Law. All companies, according to the stipulation, should be licensed by the Japanese colonial government; in case of any violation, false report, or unauthorized changes, companies were subject to closure. The underlying goal was to maintain Korea "as a simple agricultural colony and market for Japanese manufactured products."[75] During and after World War I, however, Japan emerged as a major supplier of manufactured goods for a global market; more to the point, the surplus industrial capital needed a venue in which to invest. Japan was transformed "from a debtor to a creditor nation," and private Japanese industrial investment flew to Korea and Taiwan after about 1917.[76] Paralleling the abolition of the Company Law in 1920, companies founded by Koreans increased from 39 in 1918 to 99 in 1920, 163 in 1925, and 362 in 1929.[77] Yu Ir-han possibly detected the emerging opportunities for Korean enterprise.

The growing sense of Korean nationalism in the 1920s gave Yu Ir-han a pretext to return. Oliver R. Avison (1860–1956), who was the director of the Severance Hospital, invited Yu Ir-han and his Chinese wife, Hu Meili, a pediatrician, to the missionary-run Yŏnhŭi College (Yŏnhŭi chŏnmun hakkyo) and the hospital as faculty members. Avison believed that Yu Ir-han's education in the United States, his refined sense of entrepreneurship, and Hu Meili's training in modern medicine exemplified successful careers for Korean elites, whose role would presumably be central to the future of independent Korea. Accepting the invitation, Yu Ir-han sold his share of La Choy to his partner for $250,000 and returned to Korea in 1927. A couple of Koreans with experience in customs, domestic sales marketing, and modern pharmacology helped him get settled.[78] Although biographies have mostly underscored his own accomplishments, substantial support from a group of nationalists and conservative elites made Yu Ir-han's relocation smooth.

Although invited as an educator, Yu Ir-han hoped to demonstrate successful Korean entrepreneurship. He envisioned a trading company that would first import and eventually manufacture medical supplies and miscellaneous goods, thereby improving the quality of Korean lives. Accordingly, beginning in 1926, the firm began to widely introduce insecticides, tuberculosis medicines, plasters for skin diseases, and other remedies under the Yuhan Company's brand, as well as importing sundry goods like farming tools, paints, dyes, toilet paper, cosmetics, chewing gum, and chocolates, all of which were novel and a wonder to most Koreans.[79]

THE YUHAN COMPANY'S ADVERTISEMENTS

From the beginning, the Yuhan Company heavily invested in advertising. According to Kang Han-in (1907–?), who took charge of the firm's publicity from the late 1930s to the 1940s, Yuhan spent approximately 20 percent of its total sales revenue on advertising. During the firm's heyday of 1936–40, Yuhan explored every means of publicizing, including newspapers, journals, posters, and handbills. In addition to major Korean vernacular papers—the *Korean Daily*, the *East Asian Daily*, and the *Daily News*—the Yuhan Company sponsored Manchurian, Chinese, and Japanese newspapers. All in all, the firm placed its ads in twenty different newspapers, spending ¥50,000 a month between 1936 and 1940.[80] Confirming this number, a scholar counted more than 5,173 ads by the Yuhan Company between 1927 and 1937 in a single newspaper, the *Seoul Press*. This figure implies that one or two ads were placed by the company in the newspaper every day for ten years.[81]

It is not entirely clear who designed those commercial images and copy, yet the remaining records reveal that Yu Ir-han himself directed them in the early years. Kang Han-in reported there was no specialized designer when he first joined the firm in 1936. Talented employees were trained in specific techniques of advertisement, and, during his days in the company, a couple of unofficially trained workers took charge of all the firm's advertising and public relations.[82]

As a latecomer to the market, the Yuhan Company strategically accentuated its difference from already existing medical commodities. Given the growing hold of patent medicines and Japanese domination, the firm primarily underscored its connection with the United States. The first ad for its medicine in 1928 featured quinine (*kŭmgyerap*) and anthelminthic drugs, and the major selling point was the recommendation of an American-licensed pediatrician, Yu Hu Meili, Yu Ir-han's wife.[83] Fast Dyes, the company's very first advertised item, used a similar kind of wording. With the image of the trademark willow at the center, the ad stressed the product's "United States" (*miguk*) origin and the firm's specialization in directly importing all kinds of "United States' commodities" (*migukp'um*). The word for "United States," *miguk*, was repeatedly used in the less-than-twenty-word ad.

The early advertisements' emphasis on the US connection reflects more reality than rhetoric. The Yuhan Company's advantage in the mid-1920s lay mostly in its business relationship with major Western pharmaceutical companies, such as Abbott, Parke-Davis, and Johnson & Johnson. From Abbott, for example, Yuhan imported dozens of items, including neoarsphenamine for syphilis, vitamins A and D, disinfectant, sleeping pills, expectorants, and nitromersol (metaphen). Most Western companies identified the Yuhan Company as a local branch confined to the peninsula, but Abbott designated the company as its sole agency in East Asia and provided a warehouse at the free-trade port in Dalian, China.[84] The company's connection with major US pharmaceutical companies aptly met the growing domestic demands to access Western medical goods.

As a latecomer, Yu Ir-han lacked the human resources and cultural infrastructure to efficiently compete with his rivals. However, by exploring new markets through the hospitals run by Western missionaries or by newly emerging Korean physicians of biomedicine, he quickly solidified his company's distinction with the image of novelty, science, and ethical superiority. Whereas most patent medicines evoked their connection with a millennium-old tradition of Korean medicine by maintaining the brand names of elixir (*tan*), pill (*hwan*), paste (*ko*), and water (*su*), the Yuhan Company only used foreign titles, remaining loyal to the authority of biomedicine. In contrast to the wordy advertisements of some apothecaries, Yuhan's ads selected simpler illustrations or blank space with an image of the trademark (Figure 6). Whereas most ads centered on a single brand

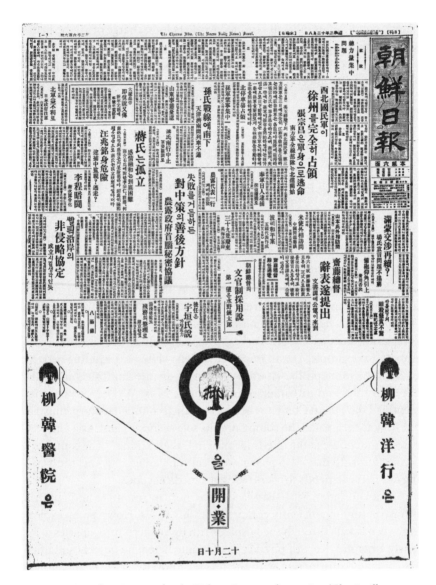

FIGURE 6. An advertisement for the Yuhan Company's opening (*Chosŏn ilbo*, December 8, 1927, p. 1)

FIGURE 7. The Yuhan Company's corporate advertisement (*Tonga ilbo*, October 30, 1938, p. 3)

item, the Yuhan Company publicized the company itself, encompassing the ethical messages of credibility, honesty, and diligence (Figure 7). Distancing themselves from patent medicines, the Yuhan advertisements hardly advised audiences to jump into self-treatment or any quick healing solution. Instead, the ads stressed that consumers should consult with doctors before they took Yuhan's medicine. In the competitive medical market in the 1920s and 1930s, Yuhan's advertising squarely identified its distinction by using the US connection, using calculated advertising techniques, and evoking a set of modern virtues for Koreans.

DIFFERENCE ENLISTED FOR
NATIONAL SURVIVAL

Every biography of Yu Ir-han cites nationalism and Christianity as the two most significant factors in shaping his successful entrepreneurship.[85] Although nationalism was surely embedded in the Yuhan Company's management, the company neither openly displayed ethnic traits nor celebrated indigenous attributes. Its ads embraced the promises of enlightenment and modernity, which often disapproved of indigenous sentiments. According to the ads, treating one's diseases meant overcoming old habits and unhygienic ways of life and eventually renewing oneself to meet the mandate of Korean modernization. The Yuhan Company framed the consumption of biomedical commodities not merely as a realm of individual desirability but as an imperative of the imagined Korean nation as a whole.

The advertisements of GU-CIDE and Neotone well exemplify this point. Determined to manufacture (not just import) biomedicines during the 1930s, Yu Ir-han sought to catch up with the latest achievements in biomedicine that might be applicable to his modern manufacturing facilities. A chemist from Vienna was invited to consult in the development of new medicine, a Korean who received a degree in Japan followed, and finally Yu Ir-han and his company were able to exploit the commercial possibilities of the recent discovery of Prontosil (Bayer).[86]

Prontosil, a derivative of a sulfanilamide (p-aminobenzene sulfonamide), was discovered in 1934 by Gerhard Domagk (1895–1964), who was awarded a Nobel Prize in Medicine in 1939. The discovery of this medicine opened new ways of commercializing the antibiotics known as sulfa drugs. For instance, Satsutarimu was the Japanese brand of related antibiotics for treating gonorrhea.[87] Not long after it was first introduced to Japan via Germany in 1937, the Yuhan Company manufactured a variety of brand items with Prontosil and sold them in domestic and overseas markets. GU-CIDE, the Korean brand, was the firm's financially successful sulfa drug (Figure 8).

FIGURE 8. An advertisement for GU-CIDE (*Chosŏn ilbo*, December 6, 1939, p. 3)

Interestingly, the success of GU-CIDE is often ascribed to its marketing catchphrase.[88] The advertisement put a great degree of emphasis on "difference": GU-CIDE was supposed to be different from any other previous medicine, as it implicitly urged readers to completely "disconnect" and "distance" themselves from the incompetence of old-style medicine and premodern inefficiency. In a sense, this catchphrase well reflected the desire of the Yuhan Company to cure Koreans' inert and sick bodies by introducing biomedicine. Inasmuch as the Yuhan Company's new medicine destroyed pathogenic bacteria, so the modernity of biomedicine would cure the deep-seated germs of Korean backwardness and inferiority. Touting the latest biomedical achievement, GU-CIDE evoked a consumption that would eventually introduce Koreans to a different mode of being.

Another of the Yuhan Company's brands was Neotone, a tonic that claimed to promote health by adding nutrition and strengthening users' immunity to disease (Figure 9). A lack of nutrition and resistance power was described not only as the weaknesses of individual bodies but as the problem of the national body as a whole. Alluding to the deterioration of the nation's health in general, Neotone encouraged people to consume medicine to be a healthy part of the Korean nation.[89]

An analogy between the individual and the nation is more explicitly expressed in the Yuhan Company's corporate advertisements. Although Korea had already lost its sovereignty in 1910, the firm advised people to contribute to the "Korean fund" for Korean business. The ads of the company in 1926 stressed,

> By business, the nation-state [*kukka*] becomes saved, and so does an individual. The nation-state that fails to develop business falls to the level of an uncivilized country and those who do not understand business become poor. Acknowledging this principle, we should be interested in business and work hard to find a means of survival. This is why we founded our company. We aim to show our reserved strength and do our best in order to establish a Korean business that is solid and perfect.[90]

When the company offered investment stock in the same year, it highlighted that it should be owned and managed only by Koreans. The ads said, "If we offer stocks for subscription to Caucasians [*paegin*], we are sure to easily collect a huge fund in a short time. Yet that would result in transferring our company to Caucasians, not Koreans."[91]

Racial contrast was employed to underscore the "Korean" initiative. As the company flourished, the exclusion of Japanese from stockholding and employment became explicit. A caveat here is that the Yuhan Company surely traded in Japanese insurance products, joining a branch of Japan Mail Shipping Line (Nippon yūsen gaisha) and Tokio Marine and Nichido Fire Insurance Company (Tokyo kaijō nichidō kasai hoken) in 1926.[92] Yet no business dealings with Japanese companies were made in 1936.[93] The Yuhan Company's business report and statement of account publicized in 1937 identified nineteen shareholders in Korea with a total of 14,725 stocks, four in China with 235, and one in Taiwan with 40, but no Japanese stockholders were known.[94] The Yuhan Company did not mind hiring non-Koreans, thus even a former Russian vice consul joined the company. But Yu Ir-han did not willingly accept Japanese employees.[95] The "Spirit of Yuhan" mission statement was established in the 1930s and repeated to employees at every morning meeting. It prioritized national pride as the most important mind-set of the company: "First, we should always work for the national health. Second, our nation [*minjok*] is not inferior to the Japanese one. We should work with a national pride. Third, the Yuhan Company does not exist for an individual. We work for the society, and [by means of the business] we should improve the economic status [of Korea]."[96]

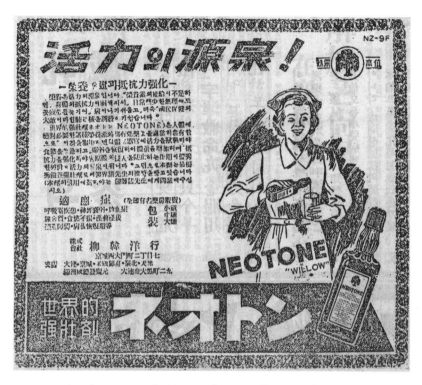

FIGURE 9. An advertisement for Neotone (*Mansŏn ilbo*, December 3, 1940, p. 3)

It is not clear whether the Yuhan Company's value-laden advertisements affected Korean consumers as intended. Yet the firm's ambitious investment in advertising paralleled its unprecedented sales records between 1926 and 1939. When the firm converted to a public limited company, the value of its gross capital reached ¥750,000. Viewed in the context of the statistics of 1938, this is half of the entire capital of thirty-three Korean pharmaceutical companies combined. The nominal capital of the largest Japanese pharmacy in Korea, Rakutendō Seiyaku, was ¥200,000 in 1933, and the most well-known Korean textile company, Keijō Textile Corporation (Kyŏngsŏng pangjik chusik hoesa), reported its paid-in capital as ¥250,000 in 1919.[97]

Under Japan's domination of capital, markets, and advertising techniques, the manufacturers of patent medicines and the Yuhan Company struggled to achieve healthy profit margins. Both desired a Korean initiative, yet they took dissimilar paths to realize their yearning. Although often piggybacking on modern worldviews, the sellers of patent medicines kept their connection with the past in terms of brand names and manufacturing methods. Elaborating on their sensitivity to the familiar flavors, local networks, and indigenous customs of Korea, the merchants of the modern "elixirs" hardly underscored national pride in health management. On the contrary, Yu Ir-han conjured up a healthy and civilized Korea through the rightful introduction of biomedicine, consumption, and industry. The nationalist pride was prioritized, yet it would be achieved by severing the old and mostly undesirable ways of life. The Yuhan Company packed its antibiotics and vitamins with the virtues of science and capitalism, which the founder passionately embraced for Korea's modernization. Here, Korea's "indigenous" faces and costumes had no role to play. It is not surprising that Neotone and GU-CIDE selectively chose Caucasian faces and Western looks for their ads.

Conclusion

The language and graphics of medical advertisements included in major Korean newspapers exhibited a prevailing mode of styling indigenous products. Against the colonial backdrop, and enthused about a growing sense of nationalism, drug sellers and manufacturers were among those who created a novel synthesis on their own terms. They also exhibited a marked sensitivity to the changing relationship between local and cosmopolitan legitimacy for their products as they negotiated the demands of their target audience. Japanese and Western agents attempting to broaden their markets in Korea were sources of new manufacturing skills and advertising strategies. Navigating the overwhelming flow of medical commodities from abroad, Korean drug sellers gradually came to establish a market for their own goods. New strategies were developed, but old styles of manufacturing and marketing continued. Consistent representation of the indigenous might not be an issue in the competitive pharmaceutical trade. However, to maximize profits under Japanese domination, Korean drug sellers had to develop a local network to attract the owners of outlying shops and use Korean manners and customs to promote sales. Early twentieth-century advertisements for medicine show that the indigenous features of Korean medicine were either made use of or ignored while their producers developed improved opportunities for Korean drug sellers.

CHAPTER FIVE

Fire Illness, or *Hwabyŏng*

Narrating Illness in the Vernacular

In the 1970s, Korean psychiatrists became puzzled by a malady commonly referred to as *hwabyŏng* (fire illness).[1] They reported that most patients who sought treatment were poor, middle-aged women with little formal education.[2] The patients had often been tormented by an abusive husband or a vicious mother-in-law for several decades and exhibited a range of psychosomatic symptoms such as anxiety, obsessive-compulsiveness, anorexia, and irritability. At first glance, Korean psychiatrists thought the symptoms similar to those of "neurotic disorders,"[3] but the American Psychiatric Association's standardized reference, the *Diagnostic and Statistical Manual of Mental Disorders* (DSM), did not appear to have any information on *hwabyŏng*.[4] Korean psychiatrists turned to their indigenous medical heritage in vain, finding that the major texts of Korean traditional medicine, such as Hŏ Chun's *Precious Mirror of Eastern Medicine* (1613) and Yi Che-ma's *Longevity and Life Preservation in Eastern Medicine* (1894), did not recognize *hwabyŏng* as a disease category.

To a certain degree, *hwabyŏng* hints at a connection with the medical heritage of Korea. The character *hwa* indicates fire and thus alludes to centuries-old discussions about "fire" in East Asian medicine.[5] In particular, Chinese physicians from the twelfth century onward began to emphasize the role of fire as an internal pathogenic agent, thereby articulating the medical understanding of human desires and emotions.[6] Sharing a textual tradition with China, Korea also adopted the Chinese articulation of fire from the sixteenth century on. Despite the Korean familiarity with discourses on fire, however, *hwabyŏng* as the label of an illness has rarely been conceptualized in the textual tradition of Korean medicine.[7]

The obscurity of *hwabyŏng* as a textual label contrasts sharply with Koreans' evident recognition of the illness in their daily lives. Every Korean seems to understand what *hwabyŏng* implies. Thus, Kim Yŏng-hwan, a dentist, politician, and former minister of science and technology, put *hwabyŏng* in his campaign slogan for the presidential election in 2012: "I'll cure the Korean people's *hwabyŏng*."[8] It is no wonder that Kim Yŏl-gyu, a scholar of Korean culture, depicts *hwabyŏng* as the country's national illness.[9] More to the point, newspaper articles regularly use the term *hwabyŏng* to depict the lamentable situation of Koreans. *Chosŏn ilbo* (Korean Daily), for instance, reported that more than 80 percent of women fighting various types of cancer are also diagnosed with *hwabyŏng*.[10] Many newspapers in 2011 reported that "Rev. Mun Yik-hwan, who is a distinguished activist for South Korea's democratization and reunification movement, died from *hwabyŏng*," quoting a politician's memorandum.[11] To be sure, *hwabyŏng* serves well as a tacit signifier of a range of Korean afflictions, which sometimes lead to death.

In the 1970s, *hwabyŏng* was first known to affect women on the economic and cultural margins of Korean society, but the illness also impacted educated women like Kim Yŏng-ju, the wife of famed resistance fighter and national poet Kim Chi-ha. Kim Yŏng-ju vividly confessed that she had suffered from *hwabyŏng* for decades before being cured by a practitioner of traditional medicine. She claims, "I was sick every day. I felt like floating without any sensation in my lower body. After taking medicine prescribed by this elderly gentleman, I felt like a clod of constrained fire [*urhwa*][12] was suddenly cast out of my chest."[13]

Besides contemporary women, the affliction has been historically associated with men. One of the most well-known, reform-minded rulers in Korean history, King Chŏngjo, is known to have perished due to an ulcer aggravated by years of *hwabyŏng*. His constant complaints about the illness are revealed in court records of the time. Ten days before his death, the king stated, "This symptom is caused by *hwabyŏng* accumulating for many years. I feel worse these days, but wasn't able to resolve this problem. I do not want to talk about anything, even important issues. I also gradually feel sick and tired of meeting you vassals. The people in this court do not fear anything, so how can fire-*ki* not be clumping in my chest?"[14] *Hwabyŏng* first appeared in Korean court records in 1603, and a series of terms implying fire-related discomforts showed up more than fifty times between the seventeenth and late nineteenth centuries.[15] Although medical texts did not explicitly recognize *hwabyŏng* as the name of an illness, past court records and the contemporary circulation of the term provide justification for its reputation as Korea's national illness.

Hwabyŏng's ostensible ubiquity in Korean society, in contrast to its relative scarcity in professional medical discourse, has provoked various efforts to officially label the problem since the 1970s. As Roy Porter points out, referring to an old saying, "A disease named is a disease half cured."[16] Increasing complaints from patients urged some psychiatrists to grapple with the validity of *hwabyŏng*. Not just Korean psychiatrists but also doctors of traditional medicine have actively joined the effort over the past thirty years to define, diagnose, and thereby cure this affliction. More to the point, the advance of depression as a global disease[17]—combined with the rising suicide rate in South Korea[18]—has encouraged Korean medical professionals to study cultural factors that may be detrimental to people's emotional wellness. This chapter begins by analyzing the process through which *hwabyŏng* draws on the traditional idea of "constrained fire" and the DSM's modern identification of "depressive disorders" to establish the term as a valid condition within professional medical circles.

Past research in Anglophone academia has rarely put *hwabyŏng* in its historical context. Most medical studies take for granted the stability of the term *hwabyŏng* and hence have traced its manifestation in accordance with the scheme of contemporary psychiatric etiology. Without fully contextualizing the illness's origins and circulation, physicians and researchers have essentialized *hwabyŏng*'s entity, then compared the illness to other psychiatric conditions for identification purposes.[19] Due to a growing interest in multiculturalism, for instance, manifestations of *hwabyŏng* among Korean Americans have been studied as part of the Asian American psychiatric experience.[20] Past studies, however, have not fully analyzed the way *hwabyŏng* has been continually transformed by new research findings and professional groups' strategies, nor have they considered the clinical implications arising from the challenges of interdisciplinary study and international recognition.

This chapter does not provide another report on *hwabyŏng*'s manifestation and treatment. Instead, I trace its origins as part of the changing mode of ethnic and indigenous portraits that Korean medical professionals use to solidify their knowledge and status and strengthen their clinical strategies. In considering the evolution of how Korean medical professionals have narrated indigenous experience from the 1970s onward, this chapter argues that the medical representation of *hwabyŏng* has been flexible, overlapping, and sometimes self-effacing, disclosing those professionals' multiple strategies at the juncture of global and domestic encounters of practices and knowledge.

From Neurosis to Anger Disorder

IDENTIFYING *HWABYŎNG*: POSSIBILITIES
AND LIMITS

Yi Si-hyŏng was the first to posit Korean culture as a key factor in determining the etiology of *hwabyŏng*.[21] Several previous authors had discussed the different ways lay people in the countryside and professionals in psychiatric clinics conceptualized *hwabyŏng*. Yi Si-hyŏng went a step further by highlighting unfavorable sociocultural conditions in South Korea at large. Based on his clinical encounters with women, Yi Si-hyŏng specifically depicted *hwabyŏng* as the sickness of an oppressed society in which marginalized women find few means to express their desires and resentment.[22]

With the notion of a distinct cultural illness now in place, Min Sŏng-gil, a psychiatrist affiliated with Yonsei University (one of the most privileged private universities in South Korea), began to search for a means to translate what Korean patients and their families reported to be *hwabyŏng* into the language of psychiatry to establish an objective diagnostic standard.[23] He began by interviewing 287 outpatients who visited psychiatric clinics in three different locations for treatment of what was then diagnosed as neurosis.[24] To conduct these interviews, he primarily relied on the third edition of the *Diagnostic Interview Schedule* (DIS-III), a tool designed by the National Institute of Mental Health of the United States and subsequently translated into Korean and validated for use with Korean patients. Min Sŏng-gil relied on a standardized interview schedule to maximize the reliability of his data. Yet he acknowledged his own modification of it by deleting what he considered to be irrelevant questions while adding others. In an attempt to garner more international recognition for his modified schedule, he also invited a US psychiatrist from the Pacific/Asian American Mental Health Research Center to appropriately train the Korean interviewers.[25]

Out of 287 patients, 56 were classified as having *hwabyŏng*, with the patients and their families agreeing with the diagnosis. In a similar fashion, 157 were identified as the non-*hwabyŏng* group. According to the diagnostic categories of the DSM-III (third edition), the *hwabyŏng* patients turned out to belong to multiple diagnostic categories, whereas the non-*hwabyŏng* patients had one or no diagnoses. The most frequently found associated diagnosis for *hwabyŏng* patients was depression (major depression and dysthymic disorder, alone or combined) and somatization disorders combined.[26] The overlapping of diagnoses was the major characteristic of *hwabyŏng*.

To support his finding, Min Sŏng-gil carried out another empirical study on the isolated Pogil Island, near the southwestern coast of Korea. Residents on the remote island work mostly in agriculture and are assumed by the researcher to have preserved their indigenous culture more successfully than did urban residents; their participation in his study was seen as complementing previous research, which mostly focused on urban centers.[27] Among 562 surveyed outpatients, Min Sŏng-gil's team was able to interview 138 individuals, identifying 34 as belonging to the *hwabyŏng* group and 67 as belonging to the non-*hwabyŏng* group. Five trained interviewers applied the DIS-III and DSM-III diagnostic standards. The research outcome supported Min Sŏng-gil's earlier research. *Hwabyŏng* patients qualified for two or more combined diagnoses. Based on the diverse interview data, Min Sŏng-gil argued that "*hwabyŏng* is a syndrome with symptoms sufficiently severe to be diagnosed as neurotic disorders," including "somatization, anxiety, and depression combined."[28]

While establishing *hwabyŏng* as a valid term in psychiatry, Min Sŏng-gil also acknowledged the weak point of his research: patients' subjective statements were the only grounds for identifying the condition.[29] Patients were often confused and inconsistent in remembering how serious their psychosomatic problems were and misunderstood terms the psychiatrists used in the interviews.[30] Some patients' testimonies slipped out of the diagnostic classification of the DSM-III. More fundamentally, he found that the cultural-linguistic barrier hampered the universal application of psychiatric terminologies. He stated,

> It is problematic that our country's psychiatrists rely solely on theories of Western medicine in treating our country's patients. Socio-cultural factors that cause our patients' mental illnesses are different from those of [Western] societies. [Our patients'] manifestation of symptoms, ways of expressing them, and methods of treatments cannot be separated from the traditional attributes of our family, society and culture. More to the point, genetic and constitutional difference may not be fully conceptualized by a Western standard. Our country is developing into a Western industrial society, and the term *hwabyŏng* is becoming less familiar to most Koreans. However,

the socio-cultural attributes of our traditional society will continue to have their repercussions beneath the rapidly changing environment on the surface.[31]

Here, Min Sŏng-gil expressed his dissatisfaction with the standard-ized psychiatric disease categories, which are North American in origin. Because he believed that modern psychiatry is too entangled in its own cultural origins, he found the identification of *hwabyŏng* to be elusive. He continued to criticize the limits of psychiatric language throughout the early 1990s.

KOREAN CULTURE AND *HWABYŎNG*

Given this predicament, Min Sŏng-gil longed to find what he considered "genuinely" Korean factors that may affect mental health. He felt that Ko-rean psychiatrists could not consider the development of symptoms, their expression, and their treatment without digging thoroughly into Korean family structure and sociocultural factors. Inasmuch as psychia-trists need to know an individual patient's deep-rooted life stories for treatment, the so-called uniquely Korean way of life should be examined by (Korean) psychiatrists to design the appropriate treatment. This par-tially explains why he became interested in *han*, the supposedly Korean way of holding grudges and experiencing anger. According to Min Sŏng-gil, *hwabyŏng* and *han* have an etiological relationship. Interviewing 146 outpatients who visited the psychiatric clinic at Yonsei Severance Hospi-tal to treat their *hwabyŏng*, he was able to question 125 about their *han*, and 117 answered that *han* is relevant to *hwabyŏng* to some degree.[32]

Unpleasant experiences in life, Min Sŏng-gil argued, cause either *hwabyŏng* or *han*. The difference mostly lies in duration and the inclusion of sociopolitical vicissitudes. Whereas *hwabyŏng* addresses comparatively recent losses and the despair of an individual, the causes of *han* stem from events experienced decades earlier, such as personal difficulties due to Japanese colonization or the Korean War (1950–53). The most frequently reported emotional reactions related to *hwabyŏng* among *han* patients are "mortification," "frustration," "nihilistic mood," "anxiety," and "regret."

Although *han* plausibly contextualizes *hwabyŏng*, validating it as a cause of *hwabyŏng* added more questions than it answered. For instance, those who have *han* without any *hwabyŏng* would need to be further examined to elaborate on this etiological relationship. The impact of stress on *han* might provide another comparative point of reference for understanding the relationship between *han* and *hwabyŏng*. The comparison between *han* and *chŏng*, another indigenous term for the Korean feeling of emotion, should also be examined. Given the need for further research, Min Sŏng-gil never hesitated to conclude that "*hwabyŏng* can be said to be a pathological condition of *han* resulting from the failure to overcome it in the long term."[33] He felt that *han* could provide an interpretative frame for understanding Koreans' collective, accumulated responses to a shared sociopolitical plight.

Obviously Min Sŏng-gil hardly doubted that *han* could speak for the nation. He viewed *han* as a uniquely Korean national (*Han'guk minjok*) expression of suffering.[34] To properly treat this problem, he called for an ethnic psychiatry that deals with the Korean "nation's neuro-psychiatry" (*minjok chŏngsin ŭihak*). His elaboration of *han* as an etiology for *hwabyŏng* resonated in later studies and continues to be quoted by psychiatrists and doctors of traditional medicine.[35] Min Sŏng-gil's research is recognized as playing a role in establishing *han* as an icon of the uniquely Korean psychological state.

The growing number of studies of *han* produced by Korean historians, theologians, and literary critics up to the 1990s have provided a substantial repertoire to support Min Sŏng-gil's argument regarding the etiological significance of *han*. *Han* is not limited to psychological or medical discourses but was also thought to represent the quintessential element of Korean aesthetics. In the 1970s, a few literary critics began to describe *han* as the most prevalent theme of Korean literary expressions. From the oldest poem to the most contemporary novel, *han* best exhibits Korea's deep-rooted emotional foundation.[36] *Minjung* theology, an indigenous Christian theology that came out of the Korean democratization movement and social activism during the 1970s and 1980s, also interpreted *han* as that which evolved from the Korean experience of the oppressed.[37] Studies on Korean music trace the origin and modification of sound that expresses *han*.[38] Inquiries into Korean shamanism never fail to point out that *han* became both the indispensable cause and the result of being drawn to indigenous spiritual experience.[39]

It is interesting to see that different avenues of Korean discourse share a dual desire to situate *han* in a specific time and place while simultaneously universalizing it. *Han* needs to be articulated through Korea's own history. For instance, Ko Ŭn, a well-known poet and thinker whose ideas represent major claims of the nationalist movement in Korean literature, put the origin of *han* in ancient Korea, in Kojosŏn (ca. 1000–108 BCE), and surveyed *han*'s enduring manifestation through the sociopolitical changes in the period of the Three Kingdoms (ca. 57 BCE–668 CE), the Koryŏ dynasty (918–1392), and the Chosŏn dynasty (1392–1910). In parallel with its historicity, however, Ko Ŭn explores *han*'s general validity. *Han* is compared with the Chinese idea of *hen*, the Mongolian feeling of *horosul*, the Manchurian expression of *krosocuka*, and the Japanese meaning of *oorami*.[40] This comparative approach, in a sense, puts *han* in a broader context by encompassing other peoples' experience of suffering. *Han* needs to go beyond Korea's geopolitical boundaries to resonate with other expressions of the oppressed. In other words, these discussions of *han* from the nonpsychiatric disciplines reveal the fact that the Korean elaboration of "Koreanness" is attended by the desire to go beyond the autochthonous. The idea of the indigenous requires a broader register for comparison and circulation.

HWABYŎNG BEYOND KOREA

Given the dual manifestation of *han* discourses, it is not surprising that the studies on the indigenous conditions for mental health were not entirely oriented toward domestic demands. Uniquely Korean narratives are inclined toward audiences in the outer world. In retrospect, Min Sŏng-gil confessed that *hwabyŏng* was welcomed more by international audiences than by domestic listeners. Psychiatrists abroad found a Korean case of a culture-bound syndrome to be meaningful.[41] In addition, he acknowledged that North American psychiatrists' research prior to his own studies actually motivated him to take the initiative to report *hwabyŏng* to global readers. Min Sŏng-gil viewed Keh-ming Lin's article published in the *American Journal of Psychiatry* in 1983 as the first international report on *hwabyŏng*.[42] "How come an American psychiatrist is reporting about Korean *hwabyŏng*?" he wondered.[43] Feeling challenged by this article, Min Sŏng-gil began to collect clinical cases about *hwabyŏng*.

Min Sŏng-gil's reputation has flourished along with his theories related to *hwabyŏng*. From the late 1990s onward, he has presented his work at numerous conferences held by international societies, such as the Association of Korean American Psychiatrists, the American Psychiatric Association, the World Psychiatric Association, the Pacific Rim College of Psychiatrists, the World Association of Social Psychiatry, and the World Congress of Cultural Psychiatry.[44]

The fourth edition of the DSM (DSM-IV) recognized *hwabyŏng* as a culture-bound syndrome in 1994. Following this entry to standardized nomenclature, the goals of Korean psychiatric research have become focused on gaining even more international recognition. Biological research, anger-related brain research, the development of a *hwabyŏng* scale, and drug tests have aimed to verify *hwabyŏng*'s autonomous existence.[45] While updating research methodology and broadening data collection, Min Sŏng-gil gradually shifted away from interview-based research. He began to put more emphasis on characterizing *hwabyŏng* as an anger disorder, which may be applicable to patients across cultural-ethnic boundaries.[46] Min Sŏng-gil's statistically sophisticated research in 2009 argued that *hwabyŏng* could be meaningfully distinguished from symptoms of major depression and should therefore be categorized under another novel name, such as an anger disorder.[47] He states, "*Hwabyŏng*, according to biological, psychological, and social models, can be viewed as a psychiatric barrier that has both emotional and somatic/behavior symptoms relevant to anger. *Hwabyŏng* can gain a name and be conceptualized in a DSM clinical category."[48]

One paradox regarding these efforts to universalize *hwabyŏng* as an anger disorder is that they undermine the uniquely Korean cultural etiology of *han*. Min Sŏng-gil has called for further investigation of attributes associated with anger that are found in many cultures, such as *ataque de nervios* among Hispanics.[49] Interestingly, Min Sŏng-gil's earlier passion to make the Koreanness of *hwabyŏng* visible has given way to a willingness to eliminate that distinction. Beginning with a desire to articulate the culture-bound attributes of *hwabyŏng* in the early 1980s, Min Sŏng-gil ended up with a culture-neutral conceptualization of *hwabyŏng* in the twenty-first century.

In his conceptualization of *hwabyŏng*, Min Sŏng-gil favorably referred to perspectives of traditional medicine. His earlier survey reveals that doctors of traditional medicine were mostly reluctant to accept *hwabyŏng*'s close relationship with factors of Korean culture, such as *han*, while unanimously believing in the autonomous status of the illness as a clinical entity. Korean psychiatrists, in contrast, paid more attention to indigenous culture as an etiological factor, but were reluctant to define *hwabyŏng* as a clinical category.[50] How did *hwabyŏng* gain its currency among doctors of traditional medicine? The discussion that follows turns to doctors of traditional medicine who adopted changing strategies to help establish *hwabyŏng* as a meaningful diagnostic category.

From *Ul* to Depression

Traditional medicine in Korea seized the initiative by appealing to post-colonial nationalist agendas. When Korea was liberated from Japan in 1945 and announced the establishment of the Republic of Korea in the South in 1948, structuring medical professions became central to institutionalizing the health care system. During the 1950s, members of the National Assembly who trained in biomedicine refused to acknowledge the efficacy of traditional medicine, condemning its lack of scientific foundation as a medical theory. Against this opinion, advocates of traditional medicine emphasized that nothing is more suited to the bodies of Koreans than traditional medicine. Advocates also claimed that traditional medicine would ultimately contribute to the national health care system by complementing biomedicine. Their claim gained support, and the National Medical Services Law (Kungmin ŭiryobŏp) approved two medical systems in 1951; this plural system survives to the present day. Based on this law, medical careers were classified in one of three categories: biomedical doctor and dentist; doctor of traditional medicine; or health care worker, nurse, or midwife.

Accreditation of educational institutions was enhanced in the 1960s. The College of Oriental Medicine (Tongyang ŭiyak taehak), which was the only institution available for a career in traditional medicine at that time, finally qualified and was raised to the status of a university in 1964 with a six-year educational program. Reflecting the increased interest in traditional medicine, more educational institutions have been established in the major cities since the 1970s. Today, eleven colleges and one graduate school of traditional medicine report a total of more than 4,400 enrolled students nationwide, and the total number of licensed doctors of traditional medicine in Korea amounted to 18,198 in 2014.[51] The status of traditional medicine in Korea has improved legally and academically since the 1960s, and most agents of traditional medicine have not felt threatened by the authority of science. Instead, they have enhanced the scientific foundation of traditional medicine by including biomedical subjects, such as anatomy, physiology, biochemistry, and genetics, in the undergraduate curriculum.

TRADITIONAL MEDICINE'S UNDERSTANDING
OF *HWABYŎNG*

In the early 1990s, *hwabyŏng* emerged as a significant research topic of traditional medicine.[52] Psychiatrists' endeavors to examine indigenous etiological factors seemingly challenged the professionals of traditional medicine.[53] In addition, traditional medicine itself became more specialized during this period, and thus its practitioners rushed to establish subspecialty disciplines, such as pediatric, gynecological, and psychiatric associations.[54] *Tongŭi sin'gyŏng chŏngsin kwahak hoeji* (Journal of Oriental neuropsychiatry), for instance, was first published in 1990. The ascendancy of depressive disorders on a global scale has motivated traditional medicine to further delve into *hwabyŏng*. According to a government survey, the prescription of antidepressant medication in South Korea increased by 52.3 percent between 2004 and 2008. Nearly a fifth of women in South Korea are reported to suffer from depression at some point in their lives, and female use of antidepressant medication was twice the male use between 2004 and 2008.[55] In parallel, the Korean Society for Depressive and Bipolar Disorders was established in 2001.[56] Traditional medicine's growing interest in *hwabyŏng* should be understood within this milieu of pressing needs to resolve emerging mental health issues in domestic and global settings. The number of articles containing the word "depression" or *hwabyŏng* in the *Journal of Oriental Neuropsychiatry* increased substantially after 2000. Following the initiative of Korean psychiatrists, the specialization of traditional medicine, and the increase in reported cases of depression, more and more clinics of traditional medicine marketed their expertise in treating *hwabyŏng*.

"Depression" and *hwabyŏng* in their Korean nomenclatures share a common character, *ul*, which has long been a significant etiological concept in traditional medicine in East Asia.[57] Translated as "constraint," *ul* primarily implies something that is stuck or unable to move or change because of a blocked *ki* movement. In East Asian medicine, the pathological condition of *ul* can frequently be a cause of fire, and this relationship between *ul* and fire is captured in another common name for fire illness, *urhwabyŏng*. As Volker Scheid demonstrates, *ul* has evolved over time and has been associated with various illness conditions.[58] Thus, *ul* cannot be succinctly reduced to any single illness or disease category. The rich textual repertoire and multiple clinical implications of *ul* actually empowered Korean doctors of traditional medicine to enter the discourse surrounding *hwabyŏng.* As latecomers to *hwabyŏng* studies, the professionals of traditional medicine primarily sought to establish their unique diagnostic and clinical approaches by fleshing out the traditional ideas about *ul* and fire.[59]

In his 1994 article coauthored with Hwang Ŭi-wan, Kim Chong-u, one of the most well-known *hwabyŏng* specialists, argued that traditional medicine provides a better answer to the four main questions about *hwabyŏng*: (1) why *hwabyŏng* is caused by upset, oppression, resentment, anger, and loathing; (2) why women are more vulnerable to *hwabyŏng*; (3) why symptoms of *hwabyŏng* seem similar to the traditional discourse on fire; and (4) why the pathogenic process involves a chronic disease.[60] While elaborating on those four issues, Kim Chong-u identified the liver as central to the pathology of *hwabyŏng.* Whereas Korean psychiatrists take the heart as the symbolic locus of *hwabyŏng,* thereby linking it to a neuropsychiatric problem, he differentiated his perspective by focusing on the liver.[61] He also explained that once the liver loses its distinctive function to regulate the activities of *ki* as the result of excessive anger and frustration, the constraint of liver-*ki* (*kan'gi ulgyŏl*) tends to occur. If the liver-*ki* is blocked, then the *ki* probably transforms into heat-fire (*yŏrhwa*), thereby leading to an upward counterflow (*kanhwa sangyŏk*) that resembles the symptoms of *hwabyŏng.* Kim Chong-u argued that this explains women's susceptibility to *hwabyŏng.* Due to women's reproductive function, the liver, spleen, kidney, and two kinds of conduits (*kyŏngmaek*), *ch'ungmaek* and *immaek*, become particularly important for women's health. When women reach menopause, both conduits become weak, the *ki* of the kidneys decreases, and the internal *ki* of fire becomes easily agitated. This is called the "emptiness of kidney and agitation of fire" (*sinhŏ hwadong*).[62]

Kim Chong-u applied other pathologies, such as "rising heart fire" (*simhwa sangyŏm*), "excessive liver fire" (*kanhwa hangsŏng*), and "flaming stomach fire" (*wihwa ch'isŏng*), to explain fire as an agent that determines other illness states.[63] He argued that fire is the *ki* of upward tendency, always drying up ŭm-*ki*, thereby creating the wind, moving the blood, and causing tumors. Furthermore, fire interacts with the heart, and the evil factor of fire disturbs the integrity of the mind-spirit (*simsin*). He viewed *hwabyŏng* as immoderate emotions that turned into excessive fire *ki*. Based on these terms and reasoning, Kim Chong-u argued that traditional medicine provides better pathological and etiological explanations for *hwabyŏng.* Unlike Min Sŏng-gil, who grounded his research in the interview scale authorized by North American psychiatrics, Kim Chong-u composed his theory of *hwabyŏng* through the language of liver-*ki*, fire, and *ul* or stagnation. To the doctor of traditional medicine, *hwabyŏng* is neither a mere syndrome nor a psychosomatic disorder. It is an illness with its own unique pathological process and hence cannot be equated with stress or hysteria.

Kim Chong-u's composition of *hwabyŏng* draws on a couple of referential sources. First, his use of the concepts of ŭm-yang and Five Phases directly refers to the Chinese canonical text *Huangdi neijing* (Yellow Emperor's inner canon) and its correlative construction of the human body. He also relies on contemporary textbooks of Korean traditional medicine, which put emphasis on the *Yellow Emperor's Inner Canon*'s holistic worldview. His understanding of fire is definitely drawn from the fourteenth-century Chinese physician Zhu Zhenheng's conceptualization of fire as an agent of inner physiology. Zhu Zhenheng's idea of nourishing ŭm to balance excess yang was highlighted by Kim Chong-u through his use of Hŏ Chun's *Precious Mirror of Eastern Medicine*, which is considered to be one of the first distinctively Korean medical texts.

The last important reference for Kim Chong-u was modern traditional Chinese medicine (TCM) textbooks from China, which are now available in Korea. Through his investigation of these textbooks, he discovered *Zhangshi yitong* (Mr. Zhang's comprehensive medicine, 1695), written by Zhang Lu (1617–99), who lived in seventeenth-century Jiangsu China. Previously, Zhang Lu had not been influential in Korea, but Kim Chong-u was able to draw on ideas from *Mr. Zhang's Comprehensive Medicine*, as outlined in TCM textbooks, to develop his understanding of *hwabyŏng*. For example, his interest in the manifestations of women's ŭm-*ki*-related diseases derived from these textbook presentations of Zhang Lu. Kim Chong-u also relied on TCM textbooks from modern China for information about fire's physiological and pathological functions. Even his use of Korean textbooks of traditional medicine published in the early 1990s brought more Chinese sources to his project because these books borrowed from medical textbooks from Taiwan and China. It was not uncommon for Korean doctors of traditional medicine during the 1990s to refer to textbooks, journal articles, and monographs from Taiwan and China.

The textual resources for a Korean construction of *hwabyŏng* thus include Korean psychiatrists' early research on *hwabyŏng*, such as Min Sŏng-gil's; passages from Chinese canonical texts, such as the *Yellow Emperor's Inner Canon*; well-known Korean compilations like *Precious*

Mirror of Eastern Medicine; and contemporary TCM views of fire as a physiological process. Korean doctors of traditional medicine, as latecomers to the specialized treatment of *hwabyŏng*, reinforced the rationale of using the "traditional" medicine by shrewdly utilizing the various resources available.

STANDARDIZING *HWABYŎNG*

Kim Chong-u's eclectic composition, however, did not last long. It became obvious that his strategic composition of the concept of *hwabyŏng* could not reach an audience outside Korea. Unlike Min Sŏng-gil, who made the "Korean" illness visible to international audiences and actively circulated his work outside Korea, Kim Chong-u worked with hardly any non-Korean psychiatrists or Chinese or Japanese doctors of traditional medicine. He recognized the weakness of articulating *hwabyŏng* solely based on the textual language of traditional medicine. *Hwabyŏng* necessitated a different kind of linguistic composition.

To overcome these limitations, Kim Chong-u turned to "objective" measurements, which he thought would eventually demonstrate traditional medicine's successful diagnosis and treatment of *hwabyŏng* and thereby gain a broader audience. He emphasizes, "No matter how diligently one may explain *hwabyŏng*, if it is not tested by biological and scientific methods, the research can hardly be widely credited."[64] He evolved toward a more advanced technique to quantify the condition. For instance, in the early 2000s, he took a position that viewed *hwabyŏng* as a phenomenon related to stress, which enabled him to measure the development of symptoms using an officially acknowledged scale.[65] Relying on and slightly modifying the Global Assessment of Recent Stress, Kim Chong-u examined forty-eight outpatients. Their answers to the assessment were analyzed by a paired-sample *t*-test. He referred to Thomas H. Holmes and Richard H. Rahe's "The Social Readjustment Rating Scale" (1967) and Camille Lloyd's "Life Events and Depressive Disorder Reviewed: Events as Predisposing Factors" (1980).[66] Kim Chong-u emulated the scale designs of psychological and psychosomatic research.

During the early 2000s, Kim Chong-u shifted away from stress toward depressive disorders. By equating *ul* with depression, he found an effective way to objectify *hwabyŏng*. His growing emphasis on a standardized diagnostic manual of traditional medicine suitably exhibits the overall trend of identifying *ul* with depression in Korea. After years of in-terdisciplinary collaboration, he was able to develop a "hwa-byung diag-nostic interview schedule (HBDIS)," which was supported by more so-phisticated statistical methods. Considered both valid and credible, his scale was approved by the Korean Psychological Association in 2004. Based on this scale, Kim Chong-u was able to make his own argument with a more confident tone.[67] Although earlier research in psychiatry de-scribed *hwabyŏng* as a combination of multiple symptoms or a strong manifestation of somatic disorders, Kim Chong-u emphasizes that more than 70 percent of his *hwabyŏng* sample patients turned out to have been diagnosed as having the DSM-IV's major depression disorder and dysthy-mic disorder.[68]

Kim Chong-u's study of *hwabyŏng* as a major depressive disorder also drew on TCM research models from China, in particular using the clini-cal standardizations of pattern recognition and treatment determination (*pyŏnjŭng nonch'i*). As Eric Karchmer has demonstrated, since the late 1950s "pattern recognition" has been central to codifying the uniqueness of Chinese medicine while providing a technique (often unacknowledged) for integrating Chinese medicine with biomedicine.[69] Korean practition-ers have also understood the significance of pattern recognition, and since 1995 they have been working on their own set of standardizations, called Oriental Medicine Standard (OMS)-prime, which was once marketed online.[70]

In his 2007 article, Kim Chong-u compared seventeen *hwabyŏng*-only patients with twenty patients from a *hwabyŏng* and major depression double diagnoses group. For this sample selection, he used standardized scales, such as the HBDIS, Structured Clinical Interview for DSM Dis-orders, OMS-prime, the Hamilton Rating Scale for Depression, and the Montgomery-Åsberg Depression Rating Scale. Kim Chong-u examined sociodemographic data and analyzed the two groups' pattern recogni-tion. He found no significant differences in the distribution of patients according to major patterns between the two groups. Examined by OMS-prime, both groups had a tendency to show "deficiency of ŭm and yang of the heart" most frequently. Based on this analysis, Kim Chong-u con-cluded that "*hwabyŏng* is a syndrome that has many different symptoms, but there is no difference between the *hwabyŏng* group and *hwabyŏng* with major depression group (double diagnosis) in terms of standardized pattern recognition." Because this research aimed to demonstrate overlaps where others (such as Min Sŏng-gil and other Korean psychiatrists) have seen differences, Kim Chong-u argued that his approach to the study of "*hwabyŏng* could be a new model for researching depression in Korea." He added that Koreans have a tendency to experience more somaticized symptoms of major depression and express the depression through the language of *hwabyŏng*.[71]

Kim Chong-u further developed his comparison between *hwabyŏng* and depression. He suggested,

> Although *hwabyŏng* and depression are often considered as a single and separate disease entity [*chilbyŏng*], if traditional medicine's method of pattern identification is applied, *hwabyŏng* and depression can be conceptualized as a series of various symptoms respectively, but not separate illness entities. This is due to traditional medicine's and Western medicine's fundamentally different approaches to disease. And this indicates that Western medicine requires complementary research on symptomatic approaches, whereas traditional medicine necessitates further understanding about a disease-centered approach.[72]

Although Kim Chong-u contrasted traditional medicine with its Western counterpart, he never posited traditional medicine as an alternative or complement to a psychiatric (scientific) approach. Instead, he strongly embraced the language of psychiatry, psychology, and statistics, thereby portraying his own version of traditional medicine that was well synchronized with the objective standardizations of biomedicine.[73]

Kim Chong-u's identification of *hwabyŏng* with depression has been further enhanced by a series of psychopharmaceutical research articles about traditional prescriptions. Published mostly after 2000, these articles equated depression with *ul*-related illness, to which *hwabyŏng* also belongs. By identifying depression with *hwabyŏng*, the articles were able to apply the experimental model of antidepressants to Korean traditional prescriptions, thereby objectively elaborating *hwabyŏng* as a medical entity. Chŏng Sŏn-yong, for instance, examined Coptis Decoction to Resolve Toxicity (*hwangnyŏn haedokt'ang*), one of the well-known prescriptions for *hwabyŏng* in South Korea. He states that "in order to treat depression, traditional medicine can rely on prescriptions that treat fire [pathologies]."[74] In addition to this research, an increasing number of articles published in the *Journal of Oriental Neuropsychiatry* since 2000 emulate the psychopharmaceutical research model for antidepressants.

Comparing Two Researchers

The foregoing analysis traces the way *hwabyŏng* has been defined, diagnosed, and treated by both Korean psychiatrists and doctors of traditional medicine since the 1970s. To make *hwabyŏng* sensible in medical discourse, professionals across fields have used various linguistic and numerical resources and have strategically modified *hwabyŏng*'s relationship with depressive disorders. Obviously *hwabyŏng*'s autonomous status as an illness entity is related to the authority of medical professionals and their career-building efforts.

Kim Chong-u began his research to differentiate traditional medicine's unique approach to *hwabyŏng* in terms of diagnosis and treatment.[75] As time went by, he relied more on quantitative methods and less on the language of traditional medicine, aiming to objectively demonstrate *hwabyŏng*'s autonomous existence. The advance of depression as a global disease stimulated traditional medicine's equation of *ul* with depressive disorders, thereby providing a link to interpret *hwabyŏng* as the Korean version of experiencing depression. Although Kim Chong-u and Min Sŏng-gil agreed that more quantitative and interdisciplinary research was indispensable to fully understanding *hwabyŏng*, they diverged in predicting the future of *hwabyŏng* in a global context. Whereas Min Sŏng-gil argued for reclassifying *hwabyŏng* as an anger disorder, and thereby differentiating it from depression,[76] Kim Chong-u was drawn to equating the illness with depression to a certain degree.

Both researchers successfully responded to the rise of cultural psychiatry, thereby making the "Korean" illness visible to international audiences. Although the idea of a uniquely Korean malady first encouraged their research, they eventually distanced themselves from this ethnic grounding. Although Min Sŏng-gil first celebrated the locally specific experience of the Korean psyche, thereby revealing the cultural bias of North American psychiatry, he eventually came to argue for a universal label for *hwabyŏng* that goes beyond Korea's cultural-geographical boundaries. In a similar vein, Kim Chong-u first delved into the millennium-old textual heritage of traditional medicine, the works of renowned Chinese physicians in the thirteenth century, centuries-old Korean medical scholarship, and modern TCM. However, he strongly desired to have *hwabyŏng* recognized by contemporary psychologists, pharmacologists, and psychiatrists; thus, he had to remake the condition into an entity that could circulate across cultural, national, and disciplinary boundaries.

What medical research about *hwabyŏng* tells us is that essentialized Korean attributes of the emotional disorder remain elusive. Medical professionals' reports on Koreanness illustrate the process through which a biography of the local has emerged and been modified, disclosing its (dis)-juncture with global trends. In spite of its unstable foundations of indigeneity, *hwabyŏng* nevertheless continues to represent "Korean" emotional disorders. A series of recent publications about *hwabyŏng* for popular audiences exposes the way experts' academic research, epitomized by Min Sŏng-gil's and Kim Chong-u's cases, is translated into nonacademic terms. Tracing the immediate and personal voices of *hwabyŏng* specialists outside academia, the next section asks, what does the continuing focus on the concept of *hwabyŏng* mean for Korean doctors? Simply put, why does *hwabyŏng* continue to be popular in contemporary Korea, given the lack of any explicit consensus among medical professionals? What possibility or limitation does the label of *hwabyŏng* create for doctors?

Toward a Better Sphere of Communication

Since 1997, almost a dozen popular texts about *hwabyŏng* have been published. Min Sŏng-gil's and Kim Chong-u's research articles were edited and published in monographs in 2009 and 2007, respectively. Additionally, since 2006, physicians of traditional medicine have introduced more than five new volumes about *hwabyŏng*. Based on clinical experience and personal endeavors, these primers aptly expose *hwabyŏng* specialists' diagnostic strategies and methods of treatments. As nonacademic publications, the volumes do not detail their sources of references. However, their (dis)similar voices enable us to view the various ways the research initiatives of Min Sŏng-gil and Kim Chong-u are emulated and modified by other medical experts, thereby allowing consideration of the function of *hwabyŏng* as a textual label in expanding or limiting the scope of contemporary management of emotional disorders. I examine four authors and their texts.

Hwang Ŭi-wan, a well-established *hwabyŏng* specialist, has treated patients for more than forty years. He served as chair of the Department of Neuro-Psychiatry at the Hospital of Traditional Medicine at Kyung Hee University and as president of the Korean Society of Oriental Neuro-psychiatry (Taehan hanbang sin'gyŏng chŏngsin kwahakhoe).[77] Hwang Ŭi-wan acknowledges that *hwabyŏng* has never been a single disease category in traditional medicine, but the term has long been a disease name in Korea.[78] Because he coauthored a couple of articles with Kim Chong-u, Hwang Ŭi-wan's definition of *hwabyŏng* and explanation of symptoms overlap with Kim Chong-u's earlier findings.[79]

Ch'oe Yŏng-jin and his coauthors begin their monograph by defining *hwabyŏng* as a "uniquely Korean disease" and then identifying its major cause as the sociocultural environment of the country.[80] Ch'oe Yŏng-jin argues that *hwabyŏng* has accompanied Korea's long history; thus, traditional medicine that has been a part of Koreans' lives for millennia is better situated to deal with the condition. As an alumnus of Kyung Hee University, one of the top schools of traditional medicine in Korea, Ch'oe Yŏng-jin shares a common institutional background with Hwang Ŭi-wan and Kim Chong-u. Accordingly, the major reasoning and rhetoric he develops for diagnosis and treatment are not radically different from those of Kim Chong-u and Hwang Ŭi-wan.

Yun Yong-sŏp, the third author, majored in pharmaceutics at Ch'ungnam University in Korea and then graduated from Nanjing University of Medicine (Nanjing zhongyiyao daxue) in China in 1997. As the Korean government acknowledges only the graduates of domestic institutions of traditional medicine for licensure, Yun Yong-sŏp is not allowed to run his own clinic in Korea. While managing his own apothecary, however, he has actively expressed his opinions through printed media and a blog. In his second book about *hwabyŏng*, Yun Yong-sŏp explains the illness by heavily relying on research about stress, strongly suggesting a holistic approach, and introducing teachings of Zen Buddhism.[81] If the roots of *hwabyŏng* and stress are associated with unfulfilled human desires, the ultimate treatment should be controlled by the mind. Interestingly, Yun Yong-sŏp often criticizes mainstream Korean traditional medicine's narrow understanding of major Chinese texts.

Last but not least, Kang Yong-wŏn introduces culturally embedded counseling as a major therapy for *hwabyŏng*. He agrees with Kim Chongu in viewing the illness as a Korean manifestation of depression. Criticizing the DSM-IV for its superficial approach to *hwabyŏng*, Kang Yong-wŏn highlights that the illness represents a collective trauma that is deeply associated with Korea's unprecedented struggle into modernity. Employing terms like *han* and *chŏng*, he illustrates that Koreans have not had enough opportunities to treat the emotional problems caused by fast economic development, a competitive educational system, and recent political changes.[82] Furthermore, the suppression of emotion signifies the fundamental problem of Korean linguistic convention—the radical division between elite language (Chinese in the past and English in the present) and vernacular. It is Kang Yong-wŏn's view that the elites in Korean history (at least since the seventeenth century) governed their people by monopolizing the written language, degrading any verbal expressions that could not be transformed into classical Chinese. Although Korean became the national language in the early twentieth century, in line with the growing nationalism and colonialism, the gap between written and spoken languages deterred Korean medical experts from fully grasping the expression of bodily complaints by lay people. Kang Yong-wŏn argues for the possibility of turning to Korean vernacular with an in-depth knowledge of its semantics and hermeneutics to recover the locally inspired technique of therapeutic colloquiality.

Authors of recent popular volumes about *hwabyŏng* share a common ground in using the illness label to foster their own clinical interventions. Almost no one denies that *hwabyŏng* is a uniquely Korean phenomenon.[83] Because it is fundamentally associated with Korean locality, Kang Yong-wŏn argues that doctors should diagnose not merely an individual patient but the entire "inventory" of Korean culture.[84] All of these views overlap with Min Sŏng-gil's earlier intention to discover the cultural root of *hwabyŏng* as a part of the Korean "nation's neuro-psychiatry."[85]

Even with the emphasis on indigenousness, however, Hwang Ŭi-wan, Ch'oe Yŏng-jin, and Yun Yong-sŏp do not specify how the unique elements of Korean culture are intertwined with the cause, manifestation, and treatment of the illness. Instead, they tacitly equate *hwabyŏng* with stress, thereby establishing a means to use theories about stress in defining *hwabyŏng* for lay audiences. For instance, Hwang Ŭi-wan portrays *hwabyŏng* as a state of excessive emotion, particularly caused by anger and anxiety. Although intended as a discussion of *hwabyŏng*, more than half of his book depicts the skillful management of stress in everyday life. Each chapter fleshes out how stress causes circulatory, respiratory, digestive, endocrine, musculoskeletal, skin, genito-urinary, and pediatric diseases. Based on this organization, Hwang Ŭi-wan's explanation competently uses a range of biomedical terminologies. Except for the brief introduction to his preferred prescription and his short comment on "constitutional medicine,"[86] no theories of traditional medicine are introduced.

The works of Ch'oe Yŏng-jin and Yun Yong-sŏp show a similar trend. Ch'oe Yŏng-jin puts both *hwabyŏng* and stress in his title and regards them as the combined root of hundreds of diseases. The organization of his book does not distinguish the symptoms and treatment of *hwabyŏng* from those of stress. By taking "stress" as a medium, Ch'oe Yŏng-jin relates *hwabyŏng* to other diseases such as high blood pressure, obesity, stroke, learning disabilities, sexual impotence, irritable bowel syndrome, and other chronic illnesses. Yun Yong-sŏp also interprets *hwabyŏng* as accumulated stress. Due to his personal inclination to Zen Buddhism, Yun Yong-sŏp's text provides more general advice to control human desires and cultivate one's mind. Unlike the other three authors, Kang Yong-wŏn views *hwabyŏng* as a Korean manifestation of depression, resonating with Kim Chong-u's interpretation. Like the other authors, Kang Yong-wŏn competently employs biomedical terms for his explanation.

In summation, recent popular texts about *hwabyŏng* are consonant with the earlier studies by Min Sŏng-gil and Kim Chong-u. All four authors used as examples here are in agreement regarding the indigenous nature of *hwabyŏng*, and they compose the symptoms and treatments of the illness by referring to the DSM-IV definition, stress studies, and depression research. The link with stress is not entirely new. Kim Chong-u intended to benefit from stress studies by emulating the measurement scale.[87] His desire to "scientize" *hwabyŏng* is also resonant with the other authors' strategic management of biomedical terms and weak elaboration of traditional medicine. Interestingly, the HBDIS, which was designed by Kim Chong-u, has not been well received by the examined authors. In addition, Kim Chong-u's argument to view *hwabyŏng* as a specifically Korean manifestation of depression has found only one advocate: Kang Yong-wŏn.

Stories to Be Told

Given this adjusted definition of *hwabyŏng* by doctors of traditional medicine, more thought should be given to the role the illness label plays in criticizing the contemporary landscape of Korean medicine. If stress and depression aptly replace the manifestation of *hwabyŏng*, why is the label still being used not only among patients but also among health specialists? What kinds of desires and complaints are associated with the contemporary use of *hwabyŏng* as a meaningful illness name?

One of the most explicit demands by the cited authors is to include patients' life narratives in a treatment process and establish a better relationship between doctor and patient. Treating *hwabyŏng* primarily implies fully delving into patients' life stories, and accordingly, the authors take patients' firsthand narratives as a meaningful starting point for treatment. For instance, Hwang Ŭi-wan begins his book by describing the cases of more than thirty patients who complained about or were diagnosed with *hwabyŏng*.[88] Although they can be viewed as general problems of family conflicts, career failures, the death of a loved one, or economic loss, every case expresses its own trajectory of life experience. To understand *hwabyŏng* more fully, Hwang Ŭi-wan implies, it is indispensable for doctors to listen to the stories carefully and to respond to them accordingly.

This technique of listening is key. According to Hwang Ŭi-wan, today's clinical environment is much too specialized and standardized, alienating patients and doctors. He states, "The hesitant patient who does not know which department s/he should go for her/his health concern exemplifies one case of human alienation. Today's clinics forget the wholeness of human beings and tend to serve as a repair shop. Besides, the more medicine develops, the more people are alienated. The clinic becomes corporatized and automatized."[89] Hwang Ŭi-wan hopes to recover a better patient-healer relationship. As a solution for *hwabyŏng*, he envisions better clinical communication that goes beyond simple questionnaires, tables, and numbers.

Yun Yong-sŏp agrees with Hwang Ŭi-wan in emphasizing the significance of dialogue between healer and patient. As Yun Yong-sŏp is not a licensed doctor of traditional medicine, the range of conversation he suggests includes more general principles of self-control, psychotherapy, and religious teachings. Interestingly, he allots almost a third of his book to introducing the "91 Stories of Zen" as a resource for soothing *hwabyŏng*.[90] Although he does not specify how the stories make an impact on treating the illness, he argues that teaching Zen is a reliable means to control stress and *hwabyŏng*. He believes the stories of illness can only be cured by another set of narratives; thus, the resonating healing power of Zen messages needs to be introduced.

Kang Yong-wŏn prioritizes narratives in elaborating his theory about *hwabyŏng*. He argues, "To sum up, our life is nothing but stories, and depression is also a story. It is clear that treatment should be a story."[91] His narrative-centered approach calls for counseling that is based on a refined understanding of Korean vernacular. Kang Yong-wŏn attempts to analyze the linguistic convention of ordinary Koreans (not elites) to develop "medicine carried out by our own language, healing of counseling by our own language."[92] According to Kang Yong-wŏn, colloquial Korean puts emphasis on verbs, not nouns, which indicates the language's full capacity to express human dynamics. Adjectives are rich; hence, the concrete details of individuality in things and people are aptly expressed. Furthermore, the use of the (non)substantive, a word order that highlights objects, the weakness of making logically connected long sentences, the development of immediacy and sensitivity, the lack of articles and prepositions, the significance of continuation, the qualitative approach, euphemism, and playfulness characterize the unique nature of colloquial Korean.[93] Of course, not all scholars would agree with this characterization of Korean. What should be noted here is his problematization of the disjointed relationship between language and experience, which reflects the failure of Korean medicine to fully come to grips with the expression of patients' suffering. To reach the root of *hwabyŏng* or depression, doctors as healers need to develop excellent listening skills with an empathetic mind. Kang Yong-wŏn insists that "intersubjectivity" between doctor and patient would help establish a more empathetic and meaningful relationship in clinical encounters.

Kang Yong-wŏn's call for vernacularization of Korean clinical language is resonant with Min Sŏng-gil's call for a Koreanized psychiatry in the late 1980s. Twenty-five years after Min Sŏng-gil's first call for indigenization, and even after Min Sŏng-gil's decision not to emphasize any ethnic attributes of *hwabyŏng*, opting for a much more universal term, "anger disorder," Kang Yong-wŏn repeats the turn toward indigenousness, thereby conjuring up the alternative in treating Koreans' emotional disorders. Both Min Sŏng-gil and Kang Yong-wŏn hope to understand the fuller and deeper meaning of *hwabyŏng*, which requires an articulation of the locally situated nature of illness and healing. To Kang Yong-wŏn, this endeavor implies the recovery of clinical colloquiality in Korean vernacular. To a certain degree, the continued circulation of *hwabyŏng* as a textual label parallels a rising demand to make patients' stories known and to establish them as a clinically meaningful repertoire for treatment. The unique, individual, and irreducible elements of patients' life experiences need to be fully acknowledged and analyzed with respect. Resonating with what Arthur Kleinman calls "mini-ethnography"[94] and Rita Charon's idea of "narrative medicine,"[95] the contemporary Korean *hwabyŏng* discourse partly serves as a means to critique existing clinical problems and to conjure up a vision of better clinical communication.

A recent report by the *Los Angeles Times* shows another hunger for making untold stories heard. Man Chul Cho, who has run a mental health program for Asians and Pacific Islanders in Los Angeles for more than twenty years, found out that Korean immigrants who call themselves *hwabyŏng* patients held on to their experiences of the Los Angeles riots in April 1992. Showing a range of symptoms such as paranoia, delusion, depression, and anger, Korean immigrants accumulated their unexplained resentment and anxiety for years. Instead of moving on, they have turned their anger inward, making it a part of their narrative of everyday experiences, and they continue calling it *hwabyŏng*.[96] Cho emphasizes the significance of counseling, the process of speaking out and listening to others, in an attempt to make the unaccountable accountable. Trained in Western psychiatry, Cho has gradually become a representative of *hwabyŏng* treatment. He insists that treating *hwabyŏng* is nothing but establishing a long-term and reliable relationship of talking and listening. As a listener, he not only has felt a responsibility to record patients' stories but also hopes to make those stories available for later generations. That is why Cho first contacted a reporter to listen to the patient stories he has kept for twenty years.

Conclusion

Contemporary medical discourses on *hwabyŏng* reveal neither the success of biomedicine nor the essence of Korean culture in treating emotional afflictions. Rather, the project of naming a uniquely Korean malady reflects a determination among medical professionals to make the indigenous meaningful, thereby guaranteeing a tool for gaining circulation and foreign recognition. A closer look at the studies by Korean doctors of traditional medicine and psychiatrists reveals how ephemeral and unstable their portrayal of the ethnic label of *hwabyŏng* has been. The idea of the "uniquely Korean" emotional illness was first conjured as a response to the rise of cultural psychiatry. Doctors of traditional medicine stressed the Korean nature of the illness, arguing that their profession's familiarity with Korean culture made them better qualified for effective treatment. However, after years of research, Korean medical professionals are disinclined to argue for the ethnic attributes of *hwabyŏng*. The Korean distinction, as an analytical concept, was favorably employed at first, only tentatively used, and then erased from the discourse. The uniquely Korean attributes of *hwabyŏng* have never been established by a clear consensus. As Roy Porter puts it, "The instability of medical terms is exacerbated by the multiplicity of illness lingos in circulation."[97] Through an analysis of the ever-changing meanings of "nervous" or "bilious" in Georgian England, Porter advises us "not to attempt to fix ideal, timeless meanings, in the manner of the French Academy, but to accept that words had their own evolutionary patterns, whose pathways, if not immutable, were themselves intelligible, especially when socially and historically contextualized."[98]

Despite its fragility, the label *hwabyŏng* continues to be used by medical professionals because the term partly helps them account for patients' life stories more fully and immediately. Embracing narratives as a therapeutic resource, a group of *hwabyŏng* specialists is criticizing mainstream medicine's lack of colloquiality and suggesting better techniques for patient-doctor dialogue. Interestingly, the label of *hwabyŏng* signifies a turn toward the indigenous initiative, then makes room for Korean doctors to engage with more general concerns of medicine, such as respecting the stories of illness and becoming better listeners. Medical professionals in South Korea, like those in other societies, are struggling with the challenges of modern medicine, such as spending less time with their patients, which leads to ignoring patient uniqueness and the decreasing role of patients as partners in treatment. In this sense, the label of *hwabyŏng* mirrors the general problems of medicine encountered by contemporary South Korean society more than it does the essentially Korean attributes of emotional illnesses.

Patients' agency, though not explored in this chapter, sets a future research agenda. More thought should be given to the way patients respond to the unstable nature of medical terminology. Given the medical professionals' increasing efforts to validate *hwabyŏng*, did patients unanimously agree with the professional definition of the illness? Or, as Porter describes, did patients consume the label *hwabyŏng* on their own terms, thereby accelerating the "irremediable instability of medical terminology"?[99] These questions go beyond the scope of this chapter, but an excerpt of a poem hints that patients are neither passive objects of medicalization nor the docile masses. Written by Yi Kyŏng-nim, a female poet born in 1947, the following poem exemplifies the patient's keen eye for psychiatry's intentional definition of *hwabyŏng* as a Korean women's illness.

Korean Women

I went to a psychiatrist to cure my anxiety.
The doctor, whose nameplate was encircled by two dragons, asked:
Are you getting on with your in-laws?
Do you have quarrels with your husband?
Are your children giving you a hard time?
No
No
No
Tilting his head, the doctor said:
Most Korean women suffer from one of these three.
Think carefully, and you'll find the cause.
No matter how hard I pondered, the answer was no.

Today I wanted to sleep with another man.
Today I wanted to get drunk.
Today I wanted to strip myself and loiter
in sunlit streets
Today I wished to loosen my hair, laugh uproariously,
and feel splendid agonies.

Which country's woman am I?[100]

Through the introduction of various themes, such as the female poet's sarcastic response to the male psychiatrist's authority and the idea of a woman's oppressed desires, the poet addresses how categorizing an illness can construct patients' collective identity. Although the psychiatrist urgently names her anxiety to treat it efficiently, the patient defies an easy naming of her fears and doubts the category of "Korean women." Medical terminologies are ever evolving, and as such, the efforts to fix the meaning of *hwabyŏng* in South Korea have been elusive. Ironically, this ambiguity regarding *hwabyŏng* possibly explains the longevity of the term in the country.

Epilogue

Medical knowledge not specific to local conditions was considered generally undesirable in premodern societies. The Hippocratic Corpus advises that "whoever would study medicine aright" should carefully observe a locale's specific mode of seasonal changes, wind variations, and water quality. The text suggests, "When, therefore, a physician comes to a district previously unknown to him, he should consider both its situation and its aspect to the winds. The effect of any town upon the health of its population varies according as it faces north or south, east or west. This is of the greatest importance."[1] The core lesson of medicine lies in the need to be acquainted with a place's specific geography and climate in light of the people's illness manifestations.

Exhibiting a similar sensitivity to local configuration, Korean scholars and physicians have longed to officially recognize the peculiar attributes of the Eastern Kingdom or (Chosŏn) Korea as an essential ground for advancing medical prowess. Scholarly endeavors thus indexed the vernacular names of neighboring herbs, imagined a place-specific lineage of medical innovation, and carefully documented the indigenous conditions of health and illness while translating them into the grammar of Western medicine. By doing so, medical elites authenticated their positions in a wider intellectual world.

While tracing the Korean identification of native characteristics of medicine, this book puts forth both micro- and macroanalyses. By focusing on a specific word's usage, each chapter adds depth and nuance to specific junctures of medical indigenization. An examination of major texts, authors, debates, advertisements, and sociointellectual connections through which each term was formulated complicates the Korean efforts at a particular moment. Chapter 1 focuses largely on the early fifteenth century, when the texts identified as being about local botanicals reached their peak. By interrogating the way local botanicals were documented, I stressed the possibilities and limitations of the vernacular nomenclature of materia medica in premodern Korea. Centering on the early seventeenth century, chapter 2 analyzes the origins of "Eastern medicine" as the label of a geographically distinctive tradition, then traces its modification and legacy in the late nineteenth and early twentieth centuries. I showed that the meaning of the East in medical compilations shifted away from balancing the regional differentiation against its northern and southern Chinese counterparts and instead sought to explicitly tame the advance of the West. Chapters 3 and 4 examine the early twentieth century as a turning point of Korean medicine, as doctors and scholars sought to decenter the authority of China, seeking alternative resources of novelty and authority from Japan and the West. It is not an exaggeration to say that under the colonial regime, the Japanese language, texts, and networks of professionals replaced those of China. Yet far from a radical cessation, the transition accompanied a degree of Korean emulation of and connection with Chinese references. Chapter 3 in particular explores the 1930s, in which the colonial alteration of medicine articulated the Chosŏn (Korean) category in "scientifically" reporting the specific attributes of Korean bodies. Chapter 5 begins with the late 1970s, when Korean psychiatrists began to contemplate "Koreanness" in a way that problematized the Anglo-American framework of psychiatry.

Each of the chapters focuses on a specific moment of "naming the local," and together illuminates a common ground underlying the history of Korean medicine; the Korean endeavors to assign a fixed nomenclature of medicine to authenticate native attributes have ultimately been unsuccessful. The five chapters together testify to how ephemeral and elusive each term was at any specific moment in evoking the particular nature of Korea as an object and subject of medical inquiry. Place-specific reasoning has been indispensable in advancing the art of medicine. However, Korean elites' documentation of local botanicals, Eastern medicine, the Chosŏn Korean body, and the indigenous fire illness often deviated unexpectedly from the original meanings and assumed readership. Moreover, the authors often participated in destabilizing the linguistic and clinical grounds of the medical terminologies.

I have ascribed the instability of Korean naming of the local to both the transitory nature of medical terminology and the inferior status of the Korean vernacular among the elites, who managed the production of medical knowledge. As Roy Porter reminds us, medical terms are always evolving, defying the mirror-like function of a signifier. Shigehisa Kuriyama also persuasively shows the limitation of words in capturing the "unknowable boundless plenitude of life's manifestation."[2] More tellingly, medical knowledge at its core reflects a tension between the particular yet simultaneously universal characteristics of human experience. As a necessary corollary, the peculiarity of a locale tends to gain significance only in comparison with a more general configuration of knowledge. The Hippocratic Corpus highlights that, in observing a place's specificity, "both those which are common to every country and those peculiar to a particular locality" need to be recognized.[3] Small wonder that Rita Charon, in her call for recovering the entirety of medicine through "narrative knowledge," underscores the significance in understanding the duality. As she argues,

> Unlike scientific knowledge or epidemiological knowledge, which tries to discover things about the natural world that are universally true or at least appear true to any observer, narrative knowledge enables one individual to understand particular events befalling another individual not as an instance of something that is universally true but as a singular and meaningful situation. Nonnarrative knowledge attempts to illuminate the universal by transcending the particular; narrative knowledge, by looking closely at individual human beings grappling with the conditions of life, attempts to illuminate the universal of the human condition by revealing the particular.[4]

Medical knowledge has hardly been anchored in the sphere of the purely autochthonous, and neither are the names of the local. The biographies of medical terms have displayed ever-changing modalities in capturing life's boundless manifestation and grappled with the particular and universal experiences of health and illness.

The Korean vernacular's subjugation to more authoritative foreign languages intensified the vulnerability of its names of indigenous medicine. It is important to understand that when the Chosŏn dynasty aimed to document locally available botanicals, names of herbs were mostly rendered in classical Chinese, using the Korean vernacular romanization as a complement; the textual composition of this documentation was enriched via the repertoire of Chinese references. Under the Japanese colonial regime, biomedical research by Koreans in the Korean vernacular failed to authenticate the "Chosŏn" Korean distinction as a valuable analytical unit. Korean researchers seemed primarily to take the initiative and cement a cultural distinction by using the "Chosŏn" category. Yet the longing for scholarly accountability often resulted in the mere replication of Japanese methodology, writing style, and conclusions. On reflection, the Korean commitment to medical research in the vernacular opened up a novel possibility in the 1930s, but only with limited validity and circulation.

The weakness of the Korean vernacular as a scholarly medium has persisted even after the liberation from Japan in 1945. For example, a Korean senior scholar of physical anthropology, in his retrospective report of the field in 1988, regretted that South Korean researchers since 1945 have largely emulated Japanese protocol in designing theory, subject, and methodology. Most of the Korean physical anthropologists, according to the scholar, found no problem in uncritically perpetuating the Japanese colonial mode of research. Except for the replacement of the "Chosŏnin" (the Japanese categorization of Koreans before 1945) with "Han'gugin" (the current South Korean unit of analysis for Koreans), the research in South Korea is almost identical to the Japanese research during the colonial period.[5] The scholar aptly points out that the dominance of Japanese as the language of advanced learning was not simply overcome by the increased amount of Korean research in the vernacular.

It is noteworthy that Korean documentation of the local conditions of medicine has displayed an outward-directed imagination of its audiences. To counterbalance the foreign, the privileged, or the authoritative, categories of the indigenous were rendered as a monolithic entity, thereby essentialized rather than elaborated for inward self-reflection. It was Chinese herbs, or *tangyak*, that the Chosŏn Korean label of "local botanicals," *hyangyak*, intended to challenge. Similarly, Hŏ Chun's labeling of Eastern medicine aimed to counter the southern and northern heritages of Chinese medicine. In the 1990s, Min Sŏng-gil confessed that his research about Korean indigenous *hwabyŏng* gained attention more among international audiences than among domestic medical experts. To be sure, the textual label of the "local" in Korean medicine served more as a representative of a supposedly genuine culture, homogeneous ethnicity, or pure nationality than as a signifier of the diversity or hybridity within. Yet as each chapter shows, the Korean passion for purity, separation, and clarity simultaneously entailed interweaving, connections, and even confusion.[6] The chapters illustrate that Korean physicians and scholars (and merchants as well) constantly posited the indigenous alongside the foreign, the particular with the universal, or the vernacular in relation to more authoritative languages, thereby creating a variety of amalgamations on their own terms.

The long-term perspective on the Korean naming of the local precipitates new questions beyond old frameworks. It is not important to decide the moment of complete separation—when Chinese medicine became Korean—or to explain what is uniquely Korean about Korean medicine. Rather, intriguing questions arise when the artificiality of naming was detected by historical actors, which led them to scrutinize the passages and relations out of which the local names were formulated. This book has tried to capture the probing of questioning minds rather than definitive, assertive convictions. Interestingly, when the ephemeral and inventive nature of Korean terminologies was contemplated, nuanced questions about the self followed. Chŏng Yag-yong shrewdly criticized the multiple names and their murky usages in Chosŏn Korea, thereby urging us to rethink the monolingual approach to the relations among words, things, and the moral ground of knowledge. Yi Sŏ-u, in a similar vein, sarcastically questioned the "true" names of materia medica in Chosŏn Korea, asking, "As false names give me a name, alas, who on earth am I?"[7]

A consonant voice is also found in the twentieth century in the doubt cast by Yi Kyŏng-nim on a male psychiatrist's artificial labeling of *hwabyŏng* as an indigenous illness suffered by Korean women. She criticizes the validity of "Korean women" as an essentialized category of illness description. Resisting a male psychiatrist's authoritative conceptualization of her anxiety, the poetess unravels her own desires, then asks provocatively, "Which country's woman am I?"[8]

To be sure, it is not merely the unequivocal enthusiasm to name and thereby fix the pure ground of "Koreanness" in medicine but the murmuring skepticism about the category of Koreanness itself over the centuries that this book has traced.

List of Characters

The following list of characters is divided into three sections: general terms, personal names, and text titles. Within each section, terms are listed in alphabetical order.

General Terms

aguk 我國
Asea taeyakpang 亞細亞大藥房

bianzheng lunzhi 辨證論治

chadan 紫檀
ch'ae 菜
chaegan 材幹
ch'aek 冊
Chahye pyŏngwŏn 慈惠病院
ch'amgirŭm 參吉音
Changburon 臟腑論
ch'angch'ul 蒼朮
ch'angp'o 菖蒲
Ch'anhwadang 贊化堂
Chappyŏng 雜病
chappyŏngak 雜病學
che 臍
Chesaengdang yakpang 濟生堂藥房
cheyakcha 劑藥者
cheyaksa 劑藥士
chibang 地方

chihwang 地黃
chilbyŏng 疾病
Ch'imgu 鍼灸
ch'imhyang 沈香
chinmaek 診脈
chinp'i 陳皮
chinyu 眞油
cho 燥
ch'o 草
chok 族
ch'ŏk 尺
ch'on 寸
ch'ŏn 千
chŏng 亭 unit of administration
chŏng 情 affection/the Korean feeling
  of emotion
chŏng 精 essence
Ch'ŏngsim pomyŏngdan 清心保命丹
Ch'ŏnil yakpang 天一藥房
ch'ŏnji un'gi 天地運氣
ch'ŏnsi 天時
Chŏnŭigam 典醫監

chosa 早死
Chōsen igakkai 朝鮮醫學會
Chōsen sōtokufu 朝鮮總督府
Chōsen sōtokufu keimukyoku 朝鮮総
督府警務局
Chosŏn ch'ongdokpu ŭiwŏn 朝鮮總督
府醫院
Chosŏn hanyak chohap 朝鮮漢藥組合
Chosŏnin 朝鮮人
Chosŏn ŭisa hyŏphoe 朝鮮醫師協會
Chosŏn yakhakhoe 朝鮮藥學會
Chosŏn yakhakkyo 朝鮮藥學校
chu 州
chuch'aek 籌策
chŭng 證
ch'ungmaek 衝脈
Ch'ungnam 忠南
Ch'ungnam ŭiyak chohap 忠南醫藥組合
ch'ungŏ 蟲魚

fuke 婦科

Guangzhou 廣州

ha 下
haenggŏm 行檢
ham 頷
han 寒 cold
han 汗 sweat
han 恨 resentment
Han'gugin 韓國人
Han'guk minjok 韓國民族
Han'guk yakchesahoe 韓國藥劑師會
Han'gŭl 한글
hansa 寒邪
Hansŏng pyŏngwŏn 漢城病院
hant'ong 寒痛
hanyŏk 寒疫
Hattori Hōkōkai 服部報公會
hen 恨
hoch'o 胡椒
homayu 胡麻油
hoyŏksŏng chirhwan 虎疫性疾患
hubak 厚朴
hwa 火
hwabyŏng 火病
Hwakch'ungnon 擴充論

Hwalmyŏngsu 活命水
hwan 丸
hwanggi 黃耆
hwanggi kyeji pujat'ang 黃耆桂枝附子
湯
hwangnyŏn haedokt'ang 黃連解毒湯
hyang 嚮 countering
hyang 鄉 local/countryside
hyangdaebu 鄉大夫
hyangdang 鄉黨
hyanggun 鄉軍
hyanggyo 鄉校
hyangmyŏng 鄉名
hyangni 鄉吏
hyangyak 鄉藥
hyangyu 香油
Hyeminsŏ 惠民署
hyŏndae ŭihak 現代醫學

idu 吏讀
immaek 任脈
in 人
Indan 仁丹
Inje taehakkyo 仁濟大學校
Inje ŭigwa taehak 仁濟醫科大學
inmul 人物
insam kyeji pujat'ang 人蔘桂枝附子湯

Jiangnan 江南
Jintan 仁丹
Jizhou 薺州

ka 家
kammo 感冒
kan 肝
kanghwal 羌活
kan'gi ulgyŏl 肝氣鬱結
kanhwa hangsŏng 肝火亢盛
kanhwa sangyŏk 肝火上逆
kasach'e 歌辭體
Keijō teikoku daigaku 京城帝国大学
ki 氣
ko 膏
kŏch'ŏ 居處
Kohōha 古方派
kok 穀
kolp'um 骨品

kongsa 攻瀉
kŏsŭngyu 苣藤油
ku 口
kukhwa 菊花
kukka 國家
kŭm 禽
Kŭmgyerap 金鷄蠟
kumi kangwalt'ang 九味羌活湯
kŭn 斤
kunggung 窮芎
kunsu 郡守
kwa 果 fruit
kwa 課 unit of a textbook
Kwallip pyŏngwŏn 官立病院
Kwallip ŭihakkyo 官立醫學教
kwŏn 卷
kwŏrŭm 厥陰
kyeji pujat'ang 桂枝附子湯
kyejit'ang 桂枝湯
kyŏlmyŏngja 決明子
kyŏn 肩
kyŏng 京
kyŏnghŏmbang 經驗方
kyŏngmaek 經脈
kyŏngnak 經絡
kyŏngnyun 經綸
Kyŏngsŏng kongjinhoe 京城共進會
Kyŏngsŏng pangjik chusik hoesa 京城
　紡織株式會社
Kyŏngsŏng ŭihak chŏnmun hakkyo 京
　城醫學專門學校
Kyŏngsŏng yakhak chŏnmun hakkyo
　京城藥學專門學校
kyou 交遇

li 里

maengmundong 麥門冬
maengmundongt'ang 麥門冬湯
maeyak 賣藥
maeyagŏpcha 賣藥業者
mahwang 麻黃
man 萬
Manchukuo 滿洲國
Manshū ika daigaku 滿洲醫科大學
Mansu paekpohwan 萬壽百補丸
men 門

miguk 美國
migukp'um 美國品
minjok 民族
minjok chŏngsin ŭihak 民族精神醫學
minjung 民衆
mok 目
muhuch'ae 武侯菜
mulmyŏng 物名
mun 門
muuch'ae 蕪尤菜
myŏngsil 名實

nabyak 臘藥
naebok 菜菖
Naegyŏng 內景
naeoe kyŏngsang chi to 內外境象之圖
Naeŭiwŏn 內醫院
Nanjing zhongyiyao daxue 南京中醫藥
　大學
Nihon denpō tsūshinsha 日本電報通信
　社
Nihon kanpō igakkai 日本漢方醫學會
Nippon yūsen gaisha 日本郵船会社

Oehyŏng 外形
ŏk 臆 chest
ŏk 億 one hundred million
onbyŏng ch'ibŏp 溫病治法
onyŏk 溫疫
osŏgak 烏犀角
Ōushūjin 歐洲人

paech'o 拜草
paegin 白人
Paek pyŏngwŏn 白病院
paekch'ae 白菜
Paekpohwan 百補丸
paeksul 白朮
paksa 博士
palp'yo hwahae chi che 發表和解之劑
Pangjung hyangyak mok ch'o pu 方中鄉
　藥目草部
pangyak 方略
panha 半夏
pi 比 unit of community
pi 鼻 nose
pi 脾 spleen (one of the five viscera)

pok 腹
pon 本
ponch'o 本草
ponmyŏng 本名
pŏp 法
poyang 保陽
pu 部
p'ung 風
p'ye 肺
p'yo 標 sign
p'yo 表 exterior
p'yŏn 篇
P'yŏngan 平安
P'yŏnghwadang 平和堂
pyŏngjŭng 病證
P'yŏngyang 平壤
pyŏnjŭng 辨證
pyŏnjŭng nonch'i 辨證論治

qi 氣

Rakutendō seiyaku 樂天堂製藥
ri 里 unit of distance
ri 裏 interior
ryŏ 閭

Sadannon 四端論
sahyang 麝香
samgyo habil 三教合一
samu 事務
sanghan 傷寒
sanghanjŭng 傷寒證
sangp'ung 傷風
sasang 四象
Sasang ŭihak 四象醫學
Satsutarimu サツタリム
Sech'ang yanghaeng 世昌洋行
sehoe 世會
Shōsōin 正倉院
siho 柴胡
sikkyŏn 識見
simbyŏng 審病
simhwa sangyŏm 心火上炎
simsin 心神
sin 神 spirit
sin 腎 kidneys (one of the five viscera)
sinhŏ hwadong 腎虛火動

Sirhak 實學
siryŏl 實熱
sŏ 署
sŏgak 犀角
sogun 俗云
sŏk 石
songch'ae 菘菜
Sŏngmyŏnnon 性命論
sŏŏl 庶孽
sŏse tongjŏm 西勢東占
sŏŭihak 西醫學
soŭm 少陰
soyang 少陽
sŏyŏk 西域
su 獸 beasts
su 水 water
sŭp 濕
suribujak 述而不作
sut'o pulbok 水土不服

taegangmun 大綱門
taegŭk 大戟
Taehan cheguk 大韓帝國
Taehan ŭiwŏn 大韓醫院
t'aeso 太少
t'aeŭm 太陰
t'aeyang 太陽
tan 丹
tang 黨
T'angaek 湯液
tanggwi 當歸
Tan'gun 檀君
tangyak 唐藥
tangyŏ 黨與
tansam 丹蔘
Tenyūdō 天佑堂
tiao 條
t'o 吐
t'ojil 土疾
Tokyo kaijō nichidō kasai hoken 東京海
    上日動火災保險
ton 旽
T'onggambu 統監府
Tongguk 東國
Tonghwa yakpang 同和藥房
tongin 東人
Tongsŏ ŭihak yŏn'guhoe 東西醫學研究會

tongŭi 東醫
tongyang 東洋
tongyang ŭihak 東洋醫學
Tongyang ŭiyak taehak 東洋醫藥大學
Torip pyŏngwŏn 道立病院
toryang 度量
tu 頭
tun 臀

uhwang 牛黃
ŭisaeng 醫生
Ŭiwŏllon 醫源論
ul 鬱
ŭm 陰
ŭmgi 陰氣
ŭmjŭng 陰症
ŭmyang 陰陽
urhwa 鬱火
urhwabyŏng 鬱火病

Wangqing 汪清
weisheng 衛生
wenbing 溫病
wihwa ch'isŏng 胃火熾盛
wiŭi 威儀
wŏn'gi 元氣

xiang 鄉
Xinjiang 新疆

yakchesa 藥劑士
yakchongsang 藥種商
Yakuhin oyobi yakuhin eigyō
   torishimari shikō kisoku 藥及藥品
   営業取締施行規則
yangban 兩班
yanggang 良薑
yangjŭng 陽證
yangmyŏng 陽明
yangsaeng 養生
yi 耳
yin 陰
Yinzhou 銀州
yo 腰
yongnoe 龍腦
yongyak 用藥
Yŏnhŭi chŏnmun hakkyo 延禧專門學校

yŏrhwa 熱火
Yuhan yanghaeng 柳韓洋行
yukkyŏng 六經
yuŭi 儒醫

zhezhong 折衷
Zhulinsi 竹林寺

Personal Names

Arai Kotarō 新井虎太郎
Arase Susumu 荒瀬進

Chang Ki-mu 張基茂
Chang Pong-yŏng 張鳳永
Chen Bangxian 陳邦賢
Cho Hŏn-yŏng 趙憲泳
Cho Kŭn-ch'ang 趙根昶
Cho Yong-man 趙容萬
Ch'oe Chae-yu 崔在裕
Ch'oe Tong 崔棟
Chŏng Kŭn-yang 鄭槿陽
Chŏng Yag-yong 丁若鏞

Feng Guozhang 馮國璋

Gong Xin 龔信

Himeno Yukio 姫野幸雄
Hŏ Chun 許浚
Huang Yuanyu 黃元御
Hwang Cha-hu 黃子厚
Hwang Ch'an 黃瓚
Hwang P'il-su 黃泌秀
Hwang To-yŏn 黃度淵

Imamura Yutaka 今村豊
Ishitoya Tsutomu 石戸谷勉
Itaru Shibata (Shibata Itaru) 柴田至

Kada Tetsuji 加田哲二
Kajimura Masayoshi 梶村正義
Kang Kyŏng-ae 姜敬愛
Kang P'il-mo 姜弼模

Kim Myŏng-sŏn 金鳴善
Kim San 金山
Kim Sŭng-mun 金勝文
Kim Tu-jong 金斗鍾
Kim Yŏng-hun 金永勳
Kodama Toshikuni 兒玉利國
Kou Zongshi 寇宗奭
Kubo Takeshi 久保武
Kwak In-sŏng 郭仁星

Li Gao 李杲
Li Yuanhong 黎元洪
Liu Hejian 劉河間
Liu Yu 陸羽

Miki Sakae 三木榮
Min Kang 閔疆
Min Pyŏng-ho 閔丙浩
Miwa Tarō 三輪太郎
Morishita Hiroshi 森下博

O Han-yŏng 吳漢泳

Paek In-je 白麟濟
Paek T'ae-sŏng 白泰星

Qibo 岐伯

Sasaki Teijirō 佐々木貞次郎
Satō Gōzō 佐藤剛藏
Sin Kil-gu 申佶求
Sŏ Chae-p'il 徐載弼
Sŏng Chu-bong 成周鳳
Su Song 蘇頌
Sugihara Tokuyuki 杉原德行

Takeda Masafusa 武田正房
Tang Zonghai 唐宗海
Tao Hongjing 陶弘景

Ueda Tsunekichi 上田常吉
Uemura Shunji 植村俊二

Wang Haogu 王好古
Wang Renan 汪訒庵
Wang Yinglin 王應隣

Wei Yilin 危亦林
Wŏn Chi-sang 元持常

Xin Xiu 慎修
Xu Jing 徐兢

Yakazu Dōmei 失數道明
Yamagishi Yūtarō 山岸祐太郎
Yan Xishan 閻錫山
Yi Che-ma 李濟馬
Yi Kyŏng-bong 李庚鳳
Yi Kyŏng-dong 李瓊仝
Yi Pyŏng-mo 李秉模
Yi Se-gyu 李世珪
Yi Sŏn-gŭn 李先根
Yi Ŭr-ho 李乙浩
Yi Yŏng-ch'un 李永春
Yoshimasu Tōdō 吉益東洞
Yu Ir-han 柳一韓
Yumoto Kyūshin 湯本求眞
Yun Chi-mi 尹知微
Yun Ch'i-wang 尹致旺
Yun Tong-ju 尹東柱

Zhang Ji 張機
Zhang Jiebin 張介賓
Zhang Lu 張璐
Zhu Gong 朱肱
Zhu Zhenheng 朱震亨

Text Titles

Aŏn kakpi 雅言覺非

Bencao beiyao 本草備要
Bencao jing jizhu 本草經集註
Bencao tujing 本草圖經
Bencao yanyi 本草衍義

Chajing 茶經
Chanke bichuan 産科秘傳
Chejung sinp'yŏn 濟衆新編
Ch'ŏnil yakpo 天一藥報
Chōsen igakkai zasshi 朝鮮醫學會雜誌

Chōsen yakuhō 朝鮮藥報
Chosŏn ŭibo 朝鮮醫報
Chosŏn ŭihakkye 朝鮮醫學界
Chosŏn wangjo sillok 朝鮮王朝實錄
Chungjo chilmunbang 中朝質問方
Chungjo chŏnsŭppang 中朝傳習方
Chungnam ŭiyak 忠南醫藥

Danxi xinfa 丹溪心法
Dashengbian 達生編

Gujin yijian 古今醫鑑

Haedong yŏksa 海東歷史
Hanbang ŭihak kangsŭpsŏ 漢方醫學講
   習書
Hanbang ŭiyakkye 漢方醫藥界
Hansŏng chubo 漢城週報
Hejifang 和劑方
Heji jufang 和劑局方
Huangdi bashiyi nanjing 黃帝八十一
   難經
Huangdi neijing 黃帝內經
Huangdi zhenjing 黃帝鍼經
Huangshi bazhong shu 黃氏八種書
Huangting jing 黃庭經
Huoren shu 活人書
Hwajebang 和劑方
Hyangyak chesaeng chipsŏngbang 鄉藥
   濟生集成方
Hyangyak chipsŏngbang 鄉藥集成方
Hyangyak hyemin kyŏnghŏmbang 鄉藥
   惠民經驗方
Hyangyak kanibang 鄉藥簡易方
Hyangyak kobang 鄉藥古方
Hyangyak kugŭppang 鄉藥救急方

Jiayou buzhu shennung bencao 嘉祐補
   註神農本草
Jingshi zhenglei beiji bencao 經史證類備
   急本草
Jingui yaolue qianzhu buzheng 金匱要略
   淺注補正

Kando ilbo 間島日報
Kanpō to kanyaku 漢方と漢藥

Keijō yakuhō 京城藥報
Kōkan igaku 皇漢医学
Koryŏsa 高麗史
Koryŏsa chŏryo 高麗史節要

Magwa hoet'ong 麻科會通
Manmong ilbo 滿蒙日報
Mansen no ikai 滿鮮之醫界
Mansen no keshōhin shōhō 滿鮮之化粧
   品商報
Mansŏn ilbo 滿鮮日報

Pal chukran mulmyŏng go 跋竹欄物
   名攷
P'alto chiriji 八道地理誌
Pangyak happ'yŏn 方藥合編
Piye paegyobang 備預百要方

Samguk sagi 三國史記
Samguk yusa 三國遺事
San Hezi hyangyakpang 三和子鄉藥方
Sanghan ryuyo 傷寒類要
Shanghanlun 傷寒論
Shanghanlun qianzhu buzheng 傷寒論
   淺注補正
Sigyŏng 詩經
Sinjip ŏŭi ch'waryobang 新集御醫撮
   要方

Taiping huimin heji jufang 太平惠民和
   劑局方
Taiping shenghui fang 太平聖惠方
Talsaeng pisŏ 達生秘書
T'ant'ojip 呑吐集
Tap chosŏn ŭimun 答朝鮮醫問
Tongnip sinmun 獨立新聞
Tongsŏ ŭihakpo 東西醫學報
Tongŭi pogam 東醫寶鑑
Tongŭi sasang sinp'yŏn 東醫四象新編
Tongŭi suse powŏn 東醫壽世保元
Tongyang ŭiyak 東洋醫藥

Ŭibang hwalt'u 醫方活套
Ŭibang yuch'wi 醫方類聚
Ŭijong sonik 醫宗損益
Ŭirim ch'waryo 醫林撮要

Xuanhe fengshi gaoli tujing 宣和奉使高
 麗圖經

Yejŏn chamnyŏng 禮典雜令
Yifang jijie 醫方集解
Yijing 易經
Yixue rumen 醫學入門
Yixue zhengzhuan 醫學正傳
Yŏksi manp'il 歷試漫筆

Zhang zhongjing fang 張仲景方
Zhang zhongjing wuzhanglun 張仲卿五
 臟論
Zhangshi yitong 張氏醫通
Zhongguo yixue da zidian 中國醫學大辭典
Zhongguo yixue shi 中國醫學史
Zhongxi huitong yijing jingyi 中西匯通
 醫經精義
Zhouli 周禮

Notes

Introduction

1. Hŏ Chun, *Tongŭi pogam*, preface, 54.
2. For the latest research on Hŏ Chun, see Sin Tong-wŏn, *Chosŏn saram Hŏ Chun*; Sin Tong-wŏn, *"Tongŭi pogam" kwa tongasia ŭihaksa*; and Kim Ho, *Hŏ Chun ŭi "Tongŭi pogam" yŏn'gu*. For a brief introduction to Hŏ Chun in English, see Hinrichs and Barnes, *Chinese Medicine and Healing*, 137–39.
3. Hŏ Chun, "chimye" 集例 in *Tongŭi pogam*.
4. Kim Ho, *Hŏ Chun ŭi "Tongŭi pogam" yŏn'gu*, 231–32.
5. Yi Ŭn-sŏng, *Sosŏl "Tongŭi pogam."*
6. Sin Tong-wŏn, *"Tongŭi pogam" kwa tongasia ŭihaksa*, 342, 345, 357–60.
7. *Sumin myojŏn*, preface. King Chŏngjo is known to have compiled the text. Only a handwritten copy is now available, at the Kyujanggak Institute for Korean Studies, Seoul National University. For the preface, see Kim Sin-gŭn, *Han'guk ŭiyaksa*, 584–85.
8. Sin Tong-wŏn, *Chosŏn saram ŭi saengno pyŏngsa*, 288–90. *Pangyak happ'yŏn* is still popular in contemporary South Korea.
9. See chapter 1 for the Korean efforts to naturalize licorice root. Kim Sŭl-gi, "Samnyŏn'gŭn kamch'o taeryang saengsan kisul kaebal," *Minjok ŭihak sinmun* (Minjok medicine news), September 6, 2012.
10. Regarding issues around locality and epistemology, see Chambers and Gillespie, "Locality in the History of Science," 221–24. Recent special issues in *Isis* and *East Asian Science, Technology, and Society* discuss the significance of identifying plural systems of knowledge to reposition the Eurocentric definition of "science" in a global setting. See, for instance, Sivasundaram, "Global Histories of Science"; Sivasundaram, "Sciences and the Global"; and Fan, "Global Turn in the History of Science." A brief summary of "localizing epistemology" is found in Nappi, *Monkey and the Inkpot*, 7–9.
11. Crossgrove, "Vernacularization of Science, Medicine, and Technology," 47.

12. Pollock, "Cosmopolitan and Vernacular in History."

13. Ibid., 596.

14. Ibid., 594.

15. Walker, "Medicines Trade in the Portuguese Atlantic World"; Tilley, "Global Histories, Vernacular Science, and African Genealogies"; Safier, "Global Knowledge on the Move"; Raj, *Relocating Modern Science*; Fan, *British Naturalists in Qing China*.

16. From a slightly different angle, Cooper articulates the inward identification of the indigenous, analyzing the rise of indigenous knowledge mapped on European local environments. Documentation of local botany and mineralogy increased in early modern Europe as a way of responding to the influx of exotic entities from other lands. Cooper, *Inventing the Indigenous*.

17. Lewis and Wigen, *Myth of Continents*, 132.

Chapter 1. Local Botanicals, or Hyangyak

1. *Sŏngjong sillok*, 1478. 10. 29.

2. Ibid.

3. *Sŏngjong sillok*, 1479. 2. 13.

4. Ibid.

5. *Hwaje* implies *Hejifang* or *Heji jufang* in Chinese, which is the shortened title of *Taiping huimin heji jufang* (Formulas of the pharmacy service for great peace and for the benefit of the people, 1107).

6. Miki, *Chōsen igakushi oyobi shippeishi*, 127.

7. Kim Tu-jong, *Han'guk ŭihaksa*, 206–207. For the most recent scholarship, see Yi Kyŏng-nok, "'Hyangyak chipsŏngbang' ŭi p'yŏnch'an."

8. *Hyang* is pronounced as *xiang* in Chinese.

9. See, for instance, the *Contemporary Chinese Dictionary*, or *Hanyu da zidian*.

10. Morohashi, *Dai kan-wa jiten*, 11:39571.

11. Han'guk chŏngsin munhwa yŏn'guwŏn, *Han'guk minjok munhwa tae paekkwa sajŏn*, 24:628; Lee, *New History of Korea*, 56, 76, 123, 188.

12. Chŏng Yag-yong, *Aŏn kakpi*. Chŏng Yag-yong criticizes the use of *hyang* as the signifier of "countryside." Rather, he argues that "Confucius resided in *hyangdang*" means he was in the capital, and when *hyang* was combined with the term "gentleman" (*hyangdaebu*), it means a gentleman of the capital. Chŏng Yag-yong lamented that this original meaning had been distorted in Chosŏn Korea for a long time.

13. *Sejong sillok*, 1423. 3. 22.

14. *Sejong sillok*, 1433. 6. 1.

15. Silla was one of the three kingdoms; the other two were Koguryŏ (37 BCE?–668 CE) and Paekche (18 BCE–660 CE). Silla took over Paekche in 660 and Koguryŏ in 668. Lee, *New History of Korea*, 36–44, 66–67.

16. Miki, *Chōsen igakushi oyobi shippeishi*, 16–17. For trading between Tang (618–907) and Silla, see Kim Tu-jong, *Han'guk ŭihaksa*, 78–79. A few records scattered in *Samguk sagi* (History of the three kingdoms, ca. 1145) and *Samguk yusa* (Memorabilia of the three kingdoms, ca. 1281–83) inform us that medical texts from Sui (581–618)

and Tang China were introduced to Korea and referred to as reliable documents in understanding materia medica before the first millennium. For bibliographical information on the texts of Chinese materia medica, see Unschuld, *Medicine in China: A History of Pharmaceutics*, and Zhuang, *Bencao yanjiu rumen*.

17. Miki, *Chōsen igakushi oyobi shippeishi*, 17–18.

18. Ibid., 18.

19. Ibid., 18–21.

20. Imported materia medica from China were prestige goods during the Silla kingdom. *History of the Three Kingdoms* reveals that Silla aristocrats were expected to make a conspicuous display of their status, which was measured in a system known as bone rank (*kolp'um*) according to sumptuary regulations. Found among the gems, fur goods, and plants forbidden to the lower ranks of the aristocracy were materia medica: black rhinoceros horn (*osŏgak*), red sandalwood (*chadan*), and *Aquilaria agallocha* (*ch'imhyang*). Most of these goods were imported from the continental western region (*sŏyŏk*), today's Xinjiang, and parts of Central Asia for aristocratic use. Kim Tu-jong, *Han'guk ŭihaksa*, 82–84.

21. Yi Kyŏng-nok, *Koryŏ sidae ŭiryo ŭi hyŏngsŏng kwa palchŏn*, 266. For instance, *Sinjip ŏŭi ch'waryobang* (Essential prescriptions reserved by court doctors, newly selected, 1226) and *Piye paegyobang* (One hundred essential prescriptions reserved, ca. 1270–90) were referred to by later authors of local botanicals in the Chosŏn dynasty.

22. Kim Tu-jong, *Han'guk ŭihaksa*, 141; Kim Sin-gŭn, *Han'guk ŭiyaksa*, 50, 57; Miki, *Chōsen igakushi oyobi shippeishi*, 68.

23. Miki, *Chōsen igakushi oyobi shippeishi*, 47–48.

24. Ibid., 55–58; Kim Tu-jong, *Han'guk ŭihaksa*, 119–24.

25. Kim Sin-gŭn, *Han'guk ŭiyaksa*, 45.

26. Miki, *Chōsen igakushi oyobi shippeishi*, 45–46, 49.

27. Kim Sin-gŭn, *Han'guk ŭiyaksa*, 46.

28. *Sinjip ŏŭi ch'waryobang*, preface, quoted in Yi Kyŏng-nok, *Koryŏ sidae ŭiryo ŭi hyŏngsŏng kwa palchŏn*, 261.

29. A part of *Piye paegyobang* is included in *Ŭibang yuch'wi* (Classified compilation of medical prescriptions, 1477). Quoted in Yi Kyŏng-nok, *Koryŏ sidae ŭiryo ŭi hyŏngsŏng kwa palchŏn*, 329. For more details about *Piye paegyobang*, see An Sang-u, "Koryŏ ŭisŏ 'Piye paegyobang' ŭi kojŭng."

30. *Hyangyak kugŭppang*. Since this text was intended as an emergency aid, each division contains a brief introduction to relevant symptoms, such as "food poisoning" in the first volume or the "miscellaneous disease of females," "the contrary attributes of medicinal drugs," and "the written formulae of experience passed from the past" in the third. For more details, see Kim Sin-gŭn, *Han'guk ŭiyaksa*, 57–59.

31. For "List of Local Botanicals," see *Hyangyak kugŭppang*. The list is also in Kim Sin-gŭn, *Han'guk ŭiyaksa*, 61–66.

32. Regarding *idu*, see Lee, *New History of Korea*, 57, and Nam P'ung-hyŏn, "Hyangyak chipsŏngbang ŭi hyangmyŏng e taehayŏ."

33. It is unknown how to read the *idu* pronunciation; hence, I put only Chinese characters as they were shown in the list.

34. Regarding these forty herbs, see Kim Sin-gŭn, *Han'guk ŭiyaksa*, 66–67.

35. Kim Tu-jong, *Han'guk ŭihaksa*, 141.

36. Miki, *Chōsen igakushi oyobi shippeishi*, 127.

37. The two nomenclatures—folk names for the list and local names for the prescriptions—reveal that two dissimilar works produced at different times were eventually included in *Hyangyak kugŭppang*. This view seems plausible, because when another list of local botanicals, in *Hyangyak chipsŏngbang*, was compiled during the early Chosŏn period, the list mainly relied on names in the prescriptions, thus continuing to adopt the "local name" as a means of emphasizing local identity but never incorporating the "folk names" from the "List of Local Botanicals." Kim Sin-gŭn, *Han'guk ŭiyaksa*, 67–68.

38. See my discussion about licorice in the following pages. The Korean attempt to naturalize licorice was not successful.

39. Kim Sin-gŭn, *Han'guk ŭiyaksa*, 67.

40. The herb trade was also carried out with other kingdoms. *Chosŏn wangjo sillok* comments on the herb trade with Japan and the Ryuku Islands, and South Asia was occasionally mentioned whenever its merchants came to trade. Trade was an official diplomatic practice authorized and regulated by the offices of the dynasty. Materia medica brought by Japanese merchants were primarily intended to meet royal demand, with any surplus authorized for use in private trade. Japanese merchants, however, preferred private to official trade, and thus they frequently hid some precious materials from royal eyes. Ibid., 241, 256–60; Kim Tu-jong, *Han'guk ŭihaksa*, 204–5.

41. Kim Sin-gŭn, *Han'guk ŭiyaksa*, 240.

42. One *kŭn* is now approximately six hundred grams for measuring medicinal herbs. Regarding Sino-Korean tributary relations and items of tributary offering, see Clark, "Sino-Korean Tributary Relations," 290–93.

43. A report to King Sejong in 1432 reveals that the Chosŏn dynasty was seeking efficient means to control the trading of items, such as musical instruments, texts, and materia medica. The king ordered a report on a possible way to regulate the trade. In 1433, an envoy to Ming was ordered to purchase various Chinese botanicals by observing the decorum of Ming China. For instance, see *Sejong sillok*, 1432. 4. 17; 1432. 10. 20; 1433. 8. 26.

44. *Koryŏsa*, 9:4a.

45. Ibid., 32:9b.

46. Ibid., 89:4a.

47. Han'guk insamsa p'yŏnch'an wiwŏnhoe, *Han'guk insamsa*, 145.

48. Tashiro, *Wakan (Waegwan)*, 125; Han'guk insamsa p'yŏnch'an wiwŏnhoe, *Han'guk insamsa*, 161–66.

49. Barnes, *Needles, Herbs, Gods, and Ghosts*, 173–81.

50. Kim Tu-jong, *Han'guk ŭihaksa*, 263. For the Sino-Korean medical exchanges, see Liang, "Chaoxian 'Yilin cuoyao' suozai zhongchao," and Liang, "Wang Yinglin yu 'Da chaoxian yi wen.'"

51. Liang, "Wang Yinglin yu 'Da chaoxian yi wen,'" 69–72.

52. For an account of encounters between Japanese and Korean physicians in the eighteenth century, see Trambaiolo, "Diplomatic Journeys and Medical Brush Talks."

53. Kim Sin-gŭn, *Han'guk ŭiyaksa*, 345–47.

54. *Sejong sillok*, 1423. 3. 22.

55. Kim Sin-gŭn, *Han'guk ŭiyaksa*, 370–74.

56. This text was eventually incorporated into *Sejong sillok chiriji* (Geographical gazette of the King Sejong's veritable records).

57. Kwŏn Pyŏng-t'ak, *Yangnyŏngsi yŏn'gu*, 53; Kim Sin-gŭn, *Han'guk ŭiyaksa*, 347–64.

58. Scholars have discussed the role the dynasty's attempt to amass botanicals from local districts in the fifteenth century played in the development of herbal drug markets after the late seventeenth century. The beginning of the herbal drug market in the seventeenth century highlights the role the state played in allocating botanicals, as "materia medica for offerings to the king," between provinces; circulating botanicals across borders and regional boundaries; and diffusing knowledge of local botanicals. For instance, Taegu, a central locus of the Kyŏngsang province in southeastern Korea, became a more significant place because of its collection of the materia medica for offerings to the king. During the seventeenth century, Taegu supplied more than 50 percent of the plant varieties offered to the king and more than 16 percent of the total volume of tributary herbs. In the eighteenth and nineteenth centuries, Taegu grew into a major private herb market center with more than one hundred permanent herb shops and thousands of agents engaged in this trade. Kwŏn Pyŏng-t'ak, *Yangnyŏngsi yŏn'gu*; Kim Tae-wŏn, "18 segi min'gan ŭiryo ŭi sŏngjang."

59. McKeown, *Modern Rise of Population*.

60. Sin Tong-wŏn does not agree with Yi T'ae-jin's methodology. For evidence of a gradual increase of population, Yi T'ae-jin analyzes 260 epitaphs from the late Koryŏ dynasty. However, Sin Tong-wŏn states that the phrase "died young (*chosa*)" can hardly be equal to "infant mortality." See Sin Tong-wŏn, "Hyangyak ŭisul i in'gu rŭl chŭngga"; Yi T'ae-jin, "Ohae wa yihae pujok ŭi chipchung"; and Yi T'ae-jin, *Ŭisul kwa in'gu kŭrigo nongŏp kisul.*

61. See Mun Chung-yang, "Sejongdae kwahak kisul." *Ŭibang yuch'wi* was first intended to pull together every known piece of medical literature, and then to put them into ninety-one major (nosological) categories. After each discussion, prescriptions from various medical texts were listed chronologically. Detailed annotations and corrections were added, which implies that in-depth knowledge of those medical texts was indispensable for this sort of grand compilation. More than 150 texts from the Han, Tang, Song, and Yuan dynasties were referenced. The complete *Ŭibang yuch'wi* is known to have amounted to 365 volumes (*kwŏn*) when it was completed in 1445, but was published in 277 volumes and 246 books (*ch'aek*) in 1477. Han'guk chŏngsin munhwa yŏn'guwŏn, *Han'guk minjok munhwa tae paekkwa sajŏn*, 17:558–59.

62. Kim Sin-gŭn, *Han'guk ŭiyaksa*, 523–32; Han'guk chŏngsin munhwa yŏn'guwŏn, *Han'guk minjok munhwa tae paekkwa sajŏn*, 24:675–76.

63. *Hyangyak chipsŏngbang*, preface, 3.

64. Ibid.

65. Ibid.

66. Ibid., 4.

67. *Sejong sillok*, 1433. 8. 27.

68. *Sejong sillok*, 1442. 2. 25.

69. *Tanjong sillok*, 1454. 2. 5; *Sŏngjong sillok*, 1488. 9. 20. King Sŏngjong's intention was not actually carried out.

70. For instance, Hŏ Chun, "T'angaekp'yŏn" 湯液篇, "ch'obu" 草部, and "kŭmbu" 金部 in *Tongŭi pogam*.

71. Kim Tu-jong, *Han'guk ŭihaksa*, 219.

72. "Ch'obu sangp'um" 草部 上品 in *Hyangyak chipsŏngbang*, seventy-eighth *kwŏn*.

73. Kim Sin-gŭn, *Han'guk ŭiyaksa*, 523.

74. For instance, Hwang Ch'an wrote *Talsaeng pisŏ* (Secret text on easy childbirth, ca. nineteenth century), combining two popular Chinese *fuke* manuals: *Dashengbian* (Treatise on easy childbirth, 1715), by a lay Buddhist, Jizhai Jushi; and *Chanke bichuan* (Transmitted secrets of women's medicine, 1786), by Zhulinsi (Bamboo Grove Monastery). Koreans not only consumed popular Chinese manuals but also edited those originals to produce new versions. Regarding the scope of circulation of Chinese *fuke* manuals, see Wu, *Reproducing Women*, ch. 2. In addition, Chŏng Yag-yong's *Magwa hoet'ong* (Mastery of measles, 1798) enlisted more than sixty references, which included the latest texts from Qing China about smallpox, measles, and pediatrics. Furthermore, Hwang To-yŏn's *Pangyak happ'yŏn* (Compendium of prescriptions, 1885) has been one of the most popular medical texts in Korea since its publication. In the preface, Hwang's son Hwang P'il-su mentions that his father was emulating Qing scholar Wang Renan's (1615–94) *Bencao beiyao* (Essentials of the materia medica, 1694) and *Yifang jijie* (Analytic collection of medical formulas, 1682).

75. Miki, *Chōsen igakushi oyobi shippeishi*, 130. The "rise and decline" narrative is found not only in Korean medicine but also in Korean science and technology. Regarding this pattern, see Yung Sik Kim, "Problems and Possibilities," and Yung Sik Kim, "'Problem of China.'"

76. Kim Tu-jong, *Han'guk ŭihaksa*, 263–4.

77. Miki, *Chōsen igakushi oyobi shippeishi*, 176.

78. This poem is included in Chŏng Yag-yong, "Hubak, modan" 厚朴牡丹, in *Aŏn kakpi*.

79. For discussions on names and naming, see Nappi, *Monkey and the Inkpot*, 55, 84–86, and Needham, *Science and Civilization in China*, 7:185–91. For Korean interest in the studies about "things and names," see Hong Yun-p'yo, "Mulmyŏnggo e taehan koch'al."

80. Hong Yun-p'yo, "Mulmyŏnggo e taehan koch'al," 167–70.

81. Chŏng Yag-yong, "Yangban" 兩班, in *Aŏn kakpi*. For the terms "name and actuality," I rely on Makeham, *Name and Actuality*. For Chŏng Yag-yong's study on materia medica, see Sin Tong-wŏn, "Yuŭi ŭi kil," 199–200.

82. Chŏng Yag-yong, *Aŏn kakpi*, preface, 15.

83. Chŏng Yag-yong, "Pal chukran mulmyŏnggo" 跋竹欄物名攷, in *Yŏyudang chŏnsŏ*, first *chip*, fourteenth *kwŏn*.

84. Cooper, *Inventing the Indigenous*, 56–57.

85. Ibid., 64.

86. Raj, *Relocating Modern Science*, 27–59.

87. Today, Danzig is known as Gdańsk, a Polish city on the Baltic coast.

88. Cooper, "Latin Words, Vernacular Worlds," 33.

89. Crossgrove, "Vernacularization of Science, Medicine, and Technology." For the quotes, see 47 and 50.

Chapter 2. Eastern Medicine, or Tongŭi

1. Hwang To-yŏn, *Ŭijong sonik*, preface, 3. For an introduction to Hwang To-yŏn's family background, see Yi Sŏn-a and Yi Si-hyŏng, "Hwang To-yŏn ŭi 'Pangyak happ'yŏn' e kwanhan yŏn'gu." Born into a family of the *yangban* elite, Hwang To-yŏn prospered as a physician managing an apothecary called the Hall of Assisting Transformation (Ch'anhwadang) in Seoul. On the strength of his reputation, he was once recommended as the private doctor of King Kojong (r. 1863–1907). He produced a series of medical books that paralleled his clinical prowess. *Ŭijong sonik* is composed of twelve volumes (*kwŏn*) and seven books (*ch'aek*) and demonstrates a mastery of encyclopedic synthesis, bringing together all available knowledge about illnesses from a range of sources. His later compilations—*Ŭibang hwalt'u* (Essential prescriptions, 1869) and *Pangyak happ'yŏn* (1885)—aimed to provide a concise and convenient primer for clinical practice. *Pangyak happ'yŏn* became a popular medical text in Korea from its first publication in 1885, and is still in circulation. Sin Tong-wŏn, *Chosŏn saram ŭi saengno pyŏngsa*, 288–91.

2. Hwang To-yŏn, *Ŭijong sonik*, preface, 5–6.

3. For recent research on Korean understanding of cold-damage disorders, see Yi Sang-wŏn, " 'Tongŭi pogam' ŭi sanghan"; Kim Nam-il, "Han'guk sanghannon yŏn'gu yaksa"; Yi Kyŏng-nok, "Chosŏn chŏn'gi 'Ŭibang yuch'wi' "; O Chae-gŭn, "Chosŏn ŭisŏ 'Hyangyak chipsŏngbang' "; and Kang Chu-bong, Kwŏn Sun-jong, and Ch'oe Chun-bae, *Imsangŭi rŭl wihan sanghannon*.

4. Yamada, *Chugoku igaku wa ikani tsukuraretaka*, 169–99. For the origins of *Shanghanlun* and different editions up to the Song dynasty, see Goldschmidt, *Evolution of Chinese Medicine*, 95–102.

5. Goldschmidt, *Evolution of Chinese Medicine*, 95–96, 154, 158–63.

6. Epler, "Concept of Disease," 10–19; Goldschmidt, *Evolution of Chinese Medicine*, 10–11.

7. Hanson, *Speaking of Epidemics in Chinese Medicine*, 16–17.

8. Yamada, *Chugoku igaku wa ikani tsukuraretaka*, 183–84.

9. Kim Nam-il, "Han'guk sanghannon yŏn'gu yaksa"; Kim Sin-gŭn, *Han'guk ŭiyaksa*, 44–46.

10. Miki, *Chōsen igakushi oyobi shippeishi*, 45.

11. Kim Sin-gŭn, *Han'guk ŭiyaksa*, 45. See chapter 1 in the section on the Koryŏ's medical texts. The Chinese search for reliable woodblock copies from Korea in 1092 aligns with Goldschmidt's explanation. He suggests that the *Treatise on Cold-Damage Disorders* was not entirely accessible to scholars and physicians by the early eleventh century, but by the end of the 1060s the Song dynasty officially intended to revive the text by placing it in mainstream medical discourse. See Goldschmidt, *Evolution of Chinese Medicine*, 100.

12. Kim Nam-il, "Han'guk sanghannon yŏn'gu yaksa." *Sanghan ryuyo* (Essentials of cold-damage types) is found among medical books for educating court doctors.

13. Subcategories include "Chehan mun" 諸寒門, "Sanghan mun" 傷寒門, "chep'ung mun" 諸風門, and "chappyŏng mun" 雜病門. For the *Ŭibang yuch'wi*'s understanding of cold-damage disorders, see Yi Kyŏng-nok, "Chosŏn chŏn'gi 'Ŭibang yuch'wi'."

14. For Hŏ Chun's compilation of *Tongŭi pogam*, see the introduction of this book.

15. Sin Tong-wŏn, *Chosŏn saram Hŏ Chun*, 156; Sin Tong-wŏn, "*Tongŭi pogam*" *kwa tongasia ŭihaksa*, 249–65.

16. Hŏ Chun, "Chappyŏng p'yŏn" 雜病篇, in *Tongŭi pogam*.

17. Sin Tong-wŏn, *Chosŏn saram Hŏ Chun*, 212; Sin Tong-wŏn, "*Tongŭi pogam*" *kwa tongasia ŭihaksa*, 306–8.

18. Sin Tong-wŏn, "*Tongŭi pogam*" *kwa tongasia ŭihaksa*, 274–79.

19. For this English translation, I referred to Mitchell, Ye, and Wiseman's version of *Shanghanlun: On Cold Damage*, 41–44.

20. Hŏ Chun, "Chappyŏng p'yŏn" 雜病篇, "Han" 寒 sang 上, "T'aeyang hyŏngjŭng yongyak" 太陽形證用藥, in *Tongŭi pogam*.

21. Hŏ Chun, "Chappyŏng p'yŏn" 雜病篇, "Han" 寒 sang 上, "Yukkyŏng p'yobon" 六經標本, in *Tongŭi pogam*.

22. Yi Sang-wŏn elaborates on Hŏ Chun's organ-centered understanding of the cold-damage disorders and its relevance to Yi Che-ma's Four Constitutions medicine. See Yi Sang-wŏn, "'Tongŭi pogam' ŭi sanghan."

23. Regarding Hŏ Chun's understanding of Daoism, see Sin Tong-wŏn, "Korean Anatomical Charts."

24. Hŏ Chun, "chimye" 集例 in *Tongŭi pogam*.

25. Chen, "Nourishing Life, Cultivation and Material Culture."

26. Sin Tong-wŏn, *Chosŏn saram Hŏ Chun*, 89–92, 180.

27. Simonis, "Mad Acts, Mad Speech," 137–42.

28. Hŏ Chun, "chimye" in *Tongŭi pogam*.

29. For Kang Myŏng-gil's family background and medical achievements, see Kim Ho, "Chŏngjodae ŭiryo chŏngch'aek," 254–56.

30. Kang Myŏng-gil, *Chejung sinp'yŏn*, preface, 3.

31. Kang Myŏng-gil, "Han" 寒 in *Chejung sinp'yŏn*, first *kwŏn*.

32. More than 90 percent of the "cold" section is composed of passages from Hŏ Chun's *Precious Mirror of Eastern Medicine*. A similar trend is found throughout the *New Compilation for Benefiting People*. See Chi Ch'ang-yŏng, "'Chejung sinp'yŏn' ŭi inyong pangsik."

33. For recent research about medical cases in Chosŏn Korea, see Yi Sŏn-a, "19 segi koch'ang chibang ŭiwŏn"; Yi Ki-bok, "Ŭian ŭro salp'yŏ ponŭn Chosŏn hugi"; and Yi Ki-bok, "18 segi ŭigwan Yi Su-gi."

34. For a brief introduction to Ŭn Su-ryong, see Yi Sŏn-a, "19 segi koch'ang chibang ŭiwŏn." Ŭn Su-ryong's cases are included in his five-book (*ch'aek*) anthology titled *T'ant'ojip* (The collection of T'ant'o). Yi Sŏn-a provides the original eleven cases in traditional Chinese and their translation into Korean.

35. Ibid., 80–81.

36. I encountered this case in Yi Ki-bok, "18 segi ŭigwan Yi Su-gi," 517. *Ton* is a unit of weighing herbs or jewelry and amounts to approximately 3.75 g.

37. Yi Sŏn-a, "19 segi koch'ang chibang ŭiwŏn," 72.

38. For Yi Che-ma's early life and family background, see Pak Sŏng-sik, "Tongmu Yi Che-ma ŭi kagye wa saengae." For the latest research on Yi Che-ma, see Yi Ki-bok, "Tongmu Yi Che-ma ŭi ŭihak sasang."

39. Yi Che-ma, "Ŭiwollon" 醫源論, in *Tongŭi suse powŏn*. I refer to Ch'oe Sŭng-hun's English translation of *Tongŭi suse powŏn*, yet have replaced his "oriental" with my "Eastern" medicine.

40. Yi Che-ma, "Ŭiwollon," in *Tongŭi suse powŏn*.

41. Yi Che-ma, "Sasangin pyŏnjŭngnon" 四象人辨證論 in *Tongŭi suse powŏn*.

42. Yi Che-ma's *Tongŭi suse powŏn* is composed of four volumes (*kwŏn*). The first volume discusses Yi Che-ma's worldview and the philosophical ground of the text, and the second and third volumes elaborate symptoms and prescriptions belonging to lesser ŭm and lesser yang people, respectively. Greater ŭm and greater yang people are treated in the fourth volume.

43. Yi Che-ma, "Soŭmin sinsuyŏl p'yoyŏl pyŏngnon" 少陰人腎受熱表熱病論 in *Tongŭi suse powŏn*.

44. Pak Sŏng-sik and Song Il-byŏng, "Sasang ŭihak ŭi haksulchŏk yŏnwŏn."

45. Yi Che-ma, "Ŭiwollon," in *Tongŭi suse powŏn*.

46. Regarding the limitations of examining Yi Che-ma's local networks, see Yi Ki-bok, "Tongmu Yi Che-ma ŭi ŭihak sasang."

47. Chŏn'guk hanŭigwa taehak sasang ŭihak kyosil, *Sasang ŭihak*, 44–45. For a possible linkage with *wenbing* studies, see Kang Chu-bong, Kwŏn Sun-jong, and Ch'oe Chun-bae, *Imsangŭi rŭl wihan sanghannon kangjwa*, 148–50.

48. See Pak Sŏng-sik, "Tongmu Yi Che-ma ŭi kagye wa saengae." Regarding the social status in Chosŏn Korea, see Hwang, *Beyond Birth*.

49. Hŏ Chun, *Tongŭi pogam*, preface, 54.

50. For privatization of medicine, see Kim Tae-wŏn, "18 segi min'gan ŭiryo ŭi sŏngjang."

51. Yi Che-ma's hometown belonged to a region that had long been discriminated against in Chosŏn Korea. The north or the northwest was a political and cultural margin, whereas families of influence and key figures in academia and politics were mostly southerners. For a discussion of marginality and regional discrimination under the Chosŏn dynasty, see Sun Joo Kim, *Marginality and Subversion in Korea*, 7–11. In particular, table 1 and table 3 of the book clearly show the less privileged position of Yi Che-ma's hometown in the Hamgyŏng province. Yi Che-ma, as an individual, was barely connected to a major network of politics, scholarly learning, and medicine. He achieved medical prowess neither by passing the state examination nor by belonging to a family of famous physicians. It is known that his self-directed study, his accumulated experience in his hometown, and a dialogue with an unknown teacher helped him succeed. Unlike Hŏ Chun, who worked as a court physician and received unconventional support from a royal family, Yi Che-ma practiced and wrote as a self-taught doctor on the margins of society.

52. For the inventive nature of Sasang ŭihak in the early twentieth century, see Sin Tong-wŏn, "Nationalistic Acceptance of Sasang Medicine."

53. For a general introduction to Wŏn Chi-sang and his text, see An Sang-u and Kim Hyŏn-gu, "Sasang ŭihak ŭi imsang ŭngyong."

54. Wŏn Chi-sang, "pŏmye 凡例, in *Tongŭi sasang sinp'yŏn*.

55. Pak Yun-jae, *Han'guk kŭndae ŭihak ŭi kiwŏn*, 304–8; Sin Tong-wŏn, "How Four Different Political Systems."

56. *Tongsŏ ŭihak yŏn'guhoe wŏlbo* 1 (1924): 74.

57. Hae San Saeng, "Yi Je-ma sasang ŭiron ch'orok" 李濟馬 四象醫論抄錄, *Tongyang ŭiyak* (Eastern medicine) 1 (1935), quoted in Sin Tong-wŏn, "Nationalistic Acceptance of Sasang Medicine," 155.

58. Sin Tong-wŏn, "Nationalistic Acceptance of Sasang Medicine."

59. The exact number of extant volumes is unknown. On the journals of traditional medicine, see Chŏng Chi-hun, "Hanŭi haksul chapchi rŭl chungsimŭro."

60. For example, see *Chosŏn ŭihakkye*, vol. 1. Nine volumes of *Chosŏn ŭihakkye* were published between March 1918 and September 1919.

61. Sŏng Chu-bong, "pŏmye," in *Hanbang ŭihak kangsŭpsŏ*.

62. Sŏng Chu-bong, *Hanbang ŭihak kangsŭpsŏ*, preface, 13.

63. Huang Yuanyu was a well-known physician during the time of Chinese emperor Qianlong (r. 1736–96). He was born into a literary family in Shandong province, north China. It is said that he changed his life goal from becoming an official to becoming a doctor after realizing the importance of medicine at the age of thirty. As a productive writer, Huang Yuanyu is known to have authored thirteen titles (*bu*), twelve of which deal with medicine. His medical theory was developed from his study about *Yijing* (The book of changes). See Liu Yanan, "Huang Yuanyu shengping ji qi heshu," 26–28.

64. Cho Hak-chun, "Ilche kangjŏmgi ŭi hanŭihak kyojae." Sŏng Chu-bong puts *Huangshi bazhong shu* (Mr. Huang's eight treatises) in his reference list, implying his reliance on eight texts authored by Huang Yuanyu. Liu Yanan, "Huang Yuanyu shengping ji qi heshu," 26.

65. For the attributes of Huang Yuanyu's clinical principles, see Liu Yanan, "Huang Yuanyu shengping ji qi heshu," 27.

66. Sin Kil-gu, "Hanŭihak kye ŭi sae kiun," in Cho Hŏn-yŏng et al., *Hanŭihak ŭi pip'an kwa haesŏl*, 205–24.

67. For instance, see *Hanbang ŭiyak* (Traditional medicine), vols. 27, 28, and 33. *Hanbang ŭiyak* was published monthly by the Association of Traditional Medicine in Ch'ungnam province (Ch'ungnam ŭiyak chohap) between 1937 and 1942. The exact dates of first and last publications are unknown. See Chŏng Chi-hun, "Hanŭi haksul chapchi rŭl chungsimŭro."

68. Daidoji, "Treating Emotion-Related Disorders," 70–71.

69. Sin Kil-gu, "Hanŭihak kye ŭi sae kiun," in Cho Hŏn-yŏng et al., *Hanŭihak ŭi pip'an kwa haesŏl*, 211.

70. A group of Korean doctors of traditional medicine recently revisited the Japanese Ancient Formula Current's approach to the *Treatise on Cold-Damage Disorders*, intending to transform the clinical competence of traditional medicine. This latest Korean interest in the cold-damage disorders created a controversy with those who have maintained a more orthodox approach to them. See Kang Kŭn-ju, "Taedam: pyŏnjŭng nonch'i nŭn hŏgu e pulgwa," *Minjok ŭihak sinmun* (Minjok medicine news), February 27, 2010, and a counterargument by Kim Nam-il, "No Yŏng-bŏm hoejangnim kke," *Minjok ŭihak sinmun*, March 22, 2010.

Chapter 3. Chosŏn Koreans

1. Kim Yŏng-hun, "Hanbang ŭihak puhŭng e taehaya," *Chosŏn ilbo*, December 31, 1934. Born in 1882, Kim Yŏng-hun exemplified a successful physician of his time. He served once as a court doctor, then opened his own clinic, Poch'un ŭiwŏn,

in northern Seoul. In 1907, Kim Yŏng-hun began to teach at Tongje ŭiwon, a short-lived government-supported educational institution for traditional medicine. Surviving through the period of Japanese colonialism, the national division (1948), the Korean War (1950–53), and South Korea's rapid modernization in the 1970s, he flourished as a revered physician and educator. All of his clinical records, which had been kept for more than sixty years, were registered as national property by the Cultural Assets Administration in 2012. An Sang-u, "Ŭich'ang ŭro parabon chŏnt'ong."

2. Rendered in Japanese, *Chōsen igakkai zasshi* evolved into the most authoritative journal of biomedicine published in Korea during 1911–42. Digitized volumes from 1913 to 1942 are available at the National Library of Korea and the Library of Seoul National University.

3. Hereafter, Chosŏn ch'ongdokpu ŭiwŏn is cited as Government-General Hospital.

4. For an account of the Japanese emulation of German medicine, see Hoi-Eun Kim, *Doctors of Empire*.

5. Mukharji, *Nationalizing the Body*, 12. For a review of scholarship, see 1–34. For Japanese control of health and diseases in Korea, see Pak Yun-jae, *Han'guk kŭndae ŭihak ŭi kiwŏn*; Yoo, *Politics of Gender in Colonial Korea*; Park, "Corporeal Colonialism"; and Sonja Kim, "Contesting Bodies." The following scholarship not only deals with "Western medicine" but also fleshes out the relationship between colonialism and medicine in Korea: Sin Tong-wŏn, "How Four Different Political Systems," and Yŏnse taehakkyo ŭihaksa yŏn'guso, *Hanŭihak, singminji rŭl alt'a.*

6. The articles concerning the debate were recently compiled and published, and in the following notes I refer to both the original edition of newspaper articles and the pages of the compiled text, Cho Hŏn-yŏng et al., *Hanŭihak ŭi pip'an kwa haesŏl.* The controversy began as doctors from both traditional medicine and biomedicine began to contribute articles to the newspapers *Chosŏn ilbo* and *Tonga ilbo* beginning in 1934. In addition to medical professionals, a couple of educated people joined the discussion, and the dialogues continued for more than five months. For the significance of this controversy in the history of Korean medicine, see Chŏn Hye-ri, "1934 nyŏn hanŭihak puhŭng nonjaeng."

7. Chang Ki-mu, "Hanbang ŭihak ŭi puhŭngch'aek," *Chosŏn ilbo*, February 16, 18, 20, 1934, in *Hanŭihak*, 27–34.

8. Chŏng Kŭn-yang, "Hanbang ŭihak puhŭng munje e taehan cheŏn," *Chosŏn ilbo*, March 9, 10, 11, 14, 1934, in *Hanŭihak*, 35–43.

9. Ibid., 36–39.

10. Chang Ki-mu, "Hanbang ŭihak puhŭng munje," *Chosŏn ilbo*, April 19, 20, 21, 22, 24, 25, 26, 27, 29, May 1, 1934, in *Hanŭihak*, 67–84.

11. Ibid., 72–74.

12. Cho Hŏn-yŏng, "Tongsŏ ŭihak ŭi pigyo pip'an ŭi p'iryo," *Chosŏn ilbo*, May 3, 4, 5, 6, 8, 9, 10, 11, 1934, in *Hanŭihak*, 85–101.

13. Yi Ŭr-ho, "Chonghap ŭihak surip ŭi chŏnje," *Chosŏn ilbo*, March 15, 16, 17, 18, 20, 21, 23, 24, 25, 27, 28, 29, 1934, in *Hanŭihak*, 45–66. See 48–51 and 62–63 for Yi Ŭr-ho's ideas about medical synthesis.

14. Sin Kil-gu, "Hanŭihak kye ŭi sae kiun," in *Hanŭihak*, 206.

15. Cho Hŏn-yŏng, "Hanbang ŭihak ŭi wigi rŭl aptugo," in *Hanŭihak*, 195.

16. Ibid., 196–97; Cho Hŏn-yŏng, "Sinŭihak ŭi palchŏn kwa hanŭihak ŭi kŭmhu," in *Hanŭihak*, 227.

17. Cho Hŏn-yŏng and other advocates of traditional medicine presented a similar awareness of the times. For instance, see Yi Ŭr-ho, "Chonghap ŭihak surip ŭi chŏnje," in *Hanŭihak*, 51–52, and Chang Ki-mu, "Hanbang ŭihak puhŭng munje," in *Hanŭihak*, 73–74, 84.

18. Advocates of traditional medicine responded defensively to the overwhelming impact of biomedicine. Yi Ŭr-ho, "Chonghap ŭihak surip ŭi chŏnje," in *Hanŭihak*, 52–53; Chang Ki-mu, "Hanbang ŭihak puhŭng munje," in *Hanŭihak*, 83; Cho Hŏn-yŏng, "Hanbang ŭihak ŭi wigi rŭl aptugo," in *Hanŭihak*, 184–85.

19. Cho Hŏn-yŏng's devotion to traditional medicine became the most salient among the participants. His public lecture, radio speech, newspaper articles, and abridged book introducing traditional medicine most eloquently presented the voice of traditional medicine during the later 1930s. His book, which was first published in 1934, was annotated and republished in 2002. The commentator values the simple layout and precise explanation of the book and believes that Cho Hŏn-yŏng's argument still presents a valid criticism of the contemporary culture of biomedicine. See Cho Hŏn-yŏng, *T'ongsok hanŭihak wŏllon*.

20. Cho Hŏn-yŏng, "Hanŭihak non e taehayŏ," *Chosŏn ilbo*, October 19, 20, 23, November 2, 3, 6, 7, 8, 20, 21, 22, 1934, in *Hanŭihak*, 125–53.

21. Sin Tong-wŏn, *Hoyŏlcha, Chosŏn ŭl sŭpkyŏk hada*, 346–50.

22. Yi Ŭr-ho, "Ŭihak kaenyŏm e taehan uri ŭi t'aedo," in *Hanŭihak*, 165.

23. See, for example, Chŏng Kŭn-yang, "Cho Hŏn-yŏng ssi ŭi hanŭihak non ŭl p'yŏngham," *Chosŏn ilbo*, July 13, 14, 15, 18, 19, 20, 22, 24, 25, 26, 27, 29, 31, August 1, 2, 3, 4, 1934, in *Hanŭihak*, 103–23. Regarding Chŏng Kŭn-yang's confidence in the scientific approach, see 109, 110, 121, and 123.

24. Cho Hŏn-yŏng, "Hanŭihak non e taehayŏ," in *Hanŭihak*, 125–27.

25. Kang P'il-mo, "Yangŭi ka pon hanŭihak," in *Hanŭihak*, 235–36.

26. Sin Kil-gu, "Hanŭihak kye ŭi sae kiun," in *Hanŭihak*, 221–22. Regarding his comments on Chŏng Kŭn-yang, see 221.

27. Cho Hŏn-yŏng, "Tongsŏ ŭihak ŭi pigyo pip'an ŭi p'iryo," in *Hanŭihak*, 98–99.

28. Chŏng Kŭn-yang, "Cho Hŏn-yŏng ssi ŭi hanŭihak non ŭl p'yŏngham," in *Hanŭihak*, 105.

29. Ibid., 108.

30. For the history of literati doctors in Korea, see Kim Nam-il, *Hanŭihak e mich'in Chosŏn ŭi chisigindŭl*.

31. Contestation and compromise between biomedicine and traditional medicine were not uniquely Korean phenomena; analogous debates took place in China and Japan. For Chinese debates on the revival of traditional medicine, see Unschuld, *Medicine in China*, 229–62, and Croizier, "Ideology of Medical Revivalism." For the Japanese case, see Kawakami, *Gendai Nihon iryōshi*, 153–61; Johnston, *Modern Epidemic*, 167–77; and Terazawa, "Gender, Knowledge, and Power," 195–219. Influential groups of traditional medicine practitioners in Japan petitioned to authorize their profession. However, the Meiji government remained firm in authorizing only biomedicine because it better served public health and military purposes.

32. Ma, "Medicine Cabinet."

33. Wales and Kim, *Song of Ariran*, 67.

34. For the "rumors" about injections and biomedicine, see Rogaski, "Vampires in Plagueland."

35. For instance, see Kim Sŭng-t'ae, "Ilbon ŭl t'onghan sŏyang ŭihak," and Kim Hyŏng-sŏk, "Hanmal Han'gugin e ŭihan sŏyang ŭihak suyong." For the number of Korean and Japanese patients treated at various Japanese-run hospitals, see Pak Yun-jae, *Han'guk kŭndae ŭihak ŭi kiwŏn*, 94, 194, 254, 256.

36. Sŏ Hong-gwan and Sin Chwa-sŏp, "Ilbon injongnon kwa Chosŏnin."

37. Rogaski, "Vampires in Plagueland," 156.

38. *Mansen no ikai* was known to be first published in the early 1920s, yet now only volumes 58 (1926) to 237 (1940) are available. Not all the articles published in the journals dealt with the ethnic nature of diseases and environment in Korea. Approximately half of the articles addressed general issues around specific symptoms, patterns of diseases, and new treatments of rampant diseases, such as syphilis and pneumonia, in line with the latest theories of biomedicine.

39. For instance, see *Mansen no ikai* 65 (1926): 73–75.

40. For instance, see *Mansen no ikai* 106 (1930): 173.

41. Yi Yŏng-ch'un's (1903–1980) case reveals how a Korean student of biomedicine established his career while publishing his research in those two journals. After graduating from the Severance School of Medicine in 1929, between 1929 and 1932 Yi Yŏng-ch'un published two articles in *Mansen no ikai*, four in *Chōsen igakkai zasshi*, and two in *Transactions of Japanese Pathological Society*. Korean-specific topics, such as Koreans' lung capacity and Korean girls' menarche, were found side by side with pieces of research that were framed without any ethnic distinctions. After Yi Yŏng-ch'un earned his doctorate with a study on the relationship between nicotine and sexual hormones in Japan in 1935, his thesis and following research continued to be published in *Mansen no ikai*, *Chōsen igakkai zasshi*, *Chosŏn ŭibo*, and *Transactions of Japanese Pathological Society* until 1945. Yi Kyu-sik, "Ilche ŭi nongch'on ch'imt'al," 129–30.

42. For instance, Sugihara Tokuyuki and Ishitoya Tsutomu first published "Chōsen no kanpō" 朝鮮の漢方 in *Mansen no ikai* 68 (1926): 17–24, then sixteen more articles with the same title were published by 89 (1928).

43. Regarding biomedicine's translation of Chinese herbs, see Lei, "From Changshan to a New Anti-malarial Drug."

44. In addition to the published articles, Imamura Tomo's encyclopedic research also demonstrates Japanese interest in Korean ginseng. Imamura Tomo, *Ninjinshi*.

45. Japanese researchers paid special attention to biomedically examining the efficacy of Korean ginseng. For instance, see Sasaki Teijirō, "Chōsen ninjin 'ekisu' no butsuri kagakuteki seishitsu" 朝鮮人蔘エキスの物理化學的性質, *Mansen no ikai* 92 (1928): 27–31. Similarly, *Chōsen igakkai zasshi*, between 1927 and 1933, presented nineteen articles about Korean ginseng's medical efficacy.

46. For Japanese exercise of hygiene in colonial Korea, see Henry, "Sanitizing Empire."

47. Pak Yun-jae, *Han'guk kŭndae ŭihak ŭi kiwŏn*, 244; for more details about the period before 1916, see 191, 232–33.

48. For the Japanese-run hospitals in Korea, see chapter 4. In their systematic studies on the colonized, Japanese scholars examined every aspect of Chosŏn Korea, including climate, topography, history, customs, linguistics, religions, and artifacts. Medicine was also a part of this project. For the Japanese ethnography of Chosŏn as self-reflection, see Atkins, *Primitive Selves*.

49. Pak Yun-jae, *Han'guk kŭndae ŭihak ŭi kiwŏn*, 245.

50. *Mansen no ikai* 86 (1928), 106 (1930), 134 (1932), 175 (1935).

51. Kodama Toshikuni, "Chōsen no mararia-byō ni tsuite" 朝鮮のマラリヤ病に就て, *Mansen no ikai* 86 (1928): 17–30.

52. For the significance of statistics in Japanese colonial governing, see Pak Myŏng-gyu, *Singmin kwŏllyŏk kwa t'onggye*.

53. For this pattern of research, see *Mansen no ikai* 60 (1926), 62 (1926), 69 (1926), 83 (1928), 86 (1928). *Chōsen igakkai zasshi* reveals a similar trend. Articles examining a variety of disease manifestations in Chosŏn Korea, such as tuberculosis, digestive troubles, and parasites, appeared during the 1910s and 1920s. A narrower unit, "Seoul," appeared later in the 1930s and 1940s.

54. Chōsen sōtokufu keimukyoku, "Chōsen ni okeru shōkōnetsu no ekigakuteki kansatsu" 朝鮮に於ける猩紅熱の疫學的觀察, *Mansen no ikai* 134 (1932): 18–26.

55. Takeda Masafusa et al., "Sikyū ganshu no Naisenjin ni okeru hikaku tōkeiteki kansatsu" 子宮癌腫の内鮮人に於ける比較統計的觀察, *Mansen no ikai* 61 (1926): 47–55.

56. Yi Sŏn-gŭn, "Keijō ni okeru shōni-chifusu no tōkeiteki kansatsu" 京城における小児チフスの統計的觀察, *Mansen no ikai* 106 (1930): 37–43.

57. For a criticism of comparative approaches, see Samman, "Limits of Classical Comparative Method."

58. Takeda et al., "Sikyū ganshu no Naisenjin ni okeru hikaku tōkeiteki kansatsu."

59. Existing scholarship has analyzed Japanese biometrics under the category of "physical anthropology." For instance, see Pak Sun-yŏng, "Ilche singmin t'ongch'i ha ŭi Chosŏn," and Pak Sun-yŏng, "Ilche singmin chuŭi wa Chosŏnin." Between 1925 and 1945, biometrics articles about research on Koreans amounted to approximately 410. Among them, around 300 pieces were published by Japanese researchers in Japanese. Most of the Japanese articles were published in medical journals, such as *Chōsen igakkai zasshi* (110) and the *Journal of Anatomy* (48). Pak Sun-yŏng, "Ilche singmin t'ongch'i ha ŭi Chosŏn," 197.

60. Miwa Tarō, "Senjin shōnen jukeisya no sokuchō to shinchō to no kankei ni tsukite" 鮮人少年受刑者の足長と身長との関係につきて, *Mansen no ikai* 62 (1926): 41–48.

61. Himeno Yukio et al., "Keijō no senmon gakkō nyūgaku shigansha no taikaku no tōkeiteki kansatsu" 京城の専門学校入学志願者の体格の統計的觀察, *Mansen no ikai* 129 (1932): 19–26.

62. Kajimura Masayoshi, "Chōsenjin rochō mōkan no tōkeiteki kansatsu" 朝鮮人顱頂毛環の統計的觀察, *Mansen no ikai* 119 (1931): 1–7; Miwa Tarō, "Chōsenjin shōnen jukeisha no bunshin ni tsuite" 朝鮮人少年受刑者の文身に就て, *Mansen no ikai* 107 (1930): 15–26; Yi Yŏng-ch'un, "Chōsenjin jogakusei no gekkei shochō ni

tsuite" 朝鮮人女学生の月経初潮に就て, *Mansen no ikai* 101 (1929): 33–38; O Han-Yŏng, "Kenkō Chōsenjin kessei karushūmu-ryō ni tsuite" 健康朝鮮人血清カルシュウム量に就て, *Mansen no ikai* 132 (1932): 27–29.

63. For instance, see articles in *Mansen no ikai* 200–201 (1937), 207 (1938), and 218 (1939). *Chōsen igakkai zasshi* exhibits a similar trend. Except for Kubo Takeshi's multivolume racial-anatomical research about "Koreans" (Chōsenjin), the number of articles about biometrics amounted to fewer than ten between 1917 and 1930. The number increased to forty-eight between 1931 and 1941.

64. Morris-Suzuki, "Debating Racial Science in Wartime Japan," 354–5.

65. Ibid., 368.

66. Kubo Takeshi, "Chōsenjin no jinshu kaibōgakuteki kenkyū" 朝鮮人の(人種)解剖學的研究, *Chōsen igakkai zasshi* 17 (1917), 24 (1919), 25 (1919), 26 (1919), 27 (1919), 28 (1920), 30 (1920), 32 (1921), 33 (1921), 34 (1921), 36 (1921), 38 (1922).

67. For details about Kubo's academic training and the German impact on establishing Japanese physical anthropology, see Hoi-Eun Kim, "Anatomically Speaking."

68. For the history of the Kyŏngsŏng ŭihak chŏnmun hakkyo, see Pak Yun-jae, *Han'guk kŭndae ŭihak ŭi kiwŏn*, 285–93.

69. Kubo was able to obtain more corpses in Korea as time went by, but most of them were patients; he complained that the body parts of his samples were not normal enough to gain an understanding of the scale of the average Korean's physique. Kubo, "Chōsenjin no jinshu kaibōgakuteki kenkyū," *Chōsen igakkai zasshi* 17 (1917): 19–23.

70. Kubo, "Chōsenjin no jinshu kaibōgakuteki kenkyū," *Chōsen igakkai zasshi* 17 (1917): 24–31.

71. Kubo, "Chōsenjin no jinshu kaibōgakuteki kenkyū," *Chōsen igakkai zasshi* 27 (1919): 11.

72. Kubo, "Chōsenjin no jinshu kaibōgakuteki kenkyū," *Chōsen igakkai zasshi* 28 (1920): 40–41; 32 (1921): 18–20, 56–59; 33 (1921): 13–17; 34 (1921); 36 (1921); 38 (1922).

73. Kubo, "Chōsenjin no jinshu kaibōgakuteki kenkyū," *Chōsen igakkai zasshi* 17 (1917): 34.

74. Kubo, "Chōsenjin no jinshu kaibōgakuteki kenkyū," *Chōsen igakkai zasshi* 25 (1919): 52; 32 (1921).

75. Arase Susumu, "Chōsenjin no taishitsu jinruigakuteki kenkyū" 朝鮮人の体質人類学的研究, *Chōsen igakkai zasshi* 24, no. 1 (1934): 60–110 (part 1), 111–53 (part 2). The journal was published as volume 19, number 1 as of January 1929. Before then, there was only the volume number.

76. Pak Sun-yŏng, "Ilche singmin t'ongch'i ha ŭi Chosŏn," 197–99.

77. Ueda Tsunekichi stayed in Korea for nineteen years. For details, see Kim Ok-chu, "Kyŏngsŏng chedae ŭihakpu ŭi ch'ejil," 196.

78. Itaru Shibata, "Brain Weight of the Korean," *American Journal of Physical Anthropology* 22, no. 1 (1936): 27–35.

79. Ibid., 30.

80. Ibid.

81. Ibid., 34.

82. Ibid., 35.

83. For instance, see Pak Sun-yŏng, "Ilche singmin chuŭi wa Chosŏnin."

84. Kim Kŭn-bae, "Ilche kangjŏmgi Chosŏnindŭl." For overall numbers of Japanese and Korean doctors between 1912 and 1923, see Pak Yun-jae, *Han'guk kŭndae ŭihak ŭi kiwŏn*, 309.

85. "Chosŏn ŭisa hyŏphoe kyuch'ik," *Chosŏn ŭibo* 3, no. 1 (1933).

86. Among the published research articles, approximately 30–50 percent used the "Chosŏn" unit of analysis.

87. Yi Se-gyu, "Ch'im kwa ku e taehaya" 침과구에대하야, *Chosŏn ŭibo* 6, no. 1 (1936): 7–10.

88. Extant volumes, including the entire list of articles, clinical reports, and essays in Korean, are digitalized and available at the Library of Seoul National University. For the addressed features, see *Chosŏn ŭibo* 1, no. 1 (1930): 1; 1, no. 2 (1931); and 2, no. 3 (1932).

89. Kwak In-sŏng, "Kŏn'gang Chosŏnin ŭi 'magŭnesyum' hamyuryang e taehaya" 건강조선인의'마그네슘'함유량에대하야, *Chosŏn ŭibo* 6 (1936): 70.

90. Kim Myŏng-sŏn, "Chosŏnin ŭi chŏngsang sikp'um chung e hamyuhan chagŭksŏng hyangnyo ŭi wiaek punbi e taehan yŏnghyang" 조선인의정상식품중에 함유한자극성향료의위액분비에대한영향, *Chosŏn ŭibo* 3, no. 4 (1933): 1–4.

91. Yi Yŏng-ch'un and Ch'oe Chae-yu, "Kŏn'gang Chosŏnin ŭi hyŏltangnyang e taehaya" 건강조선인의혈당량에대하야, *Chosŏn ŭibo* 3 (1933): 66–70.

92. The term "healthy" was employed here to normalize the sample data, yet was used without any explicit definition. It merely implied people who did not have any obvious disease.

93. Kwak In-sŏng, "Kŏn'gang Chosŏnin ŭi 'magŭnesyum' hamyuryang e taehaya," *Chosŏn ŭibo* 6 (1936): 70.

94. Yi Yŏng-ch'un and Ch'oe Chae-yu, "Kŏn'gang Chosŏnin ŭi hyŏltangnyang e taehaya," *Chosŏn ŭibo* 3 (1933): 66–70.

95. Kwak In-sŏng, "Kŏngang Chosŏnin ŭi 'magŭnesyum' hamyuryang e taehaya," 70.

96. Paek T'ae-sŏng, "Chosŏnin kwan'gol ŭi injong haebuhakchŏk kwanch'al" 조선인관골의인종해부학적관찰, *Chosŏn ŭibo* 1, no. 1 (1931): 5–8.

97. Yi Sŏn-gŭn, "Chosŏnin kwa Ilbonin ŭi p'oyua ŭi saengnijŏk chosŏn e taehaya" 조선인과일본인의포유아의생리적조선에대하야, *Chosŏn ŭibo* 1, no. 3 (1932): 1–7.

98. Ch'oe Tong, "Chosŏnin ŭi kisaengch'ung" 조선의기생충, *Chosŏn ŭibo* 1, no. 1 (1930): 30–32.

99. Yun Ch'i-wang, "Chosŏnin ŭi wolgyŏng ch'ocho yŏllŏng" 조선인의월경초조 연령, *Chosŏn ŭibo* 2, no. 3 (1932): 9–10.

100. Ch'oe Tong, "Chosŏnin ŭi hyŏlhyŏng" 조선인의혈형, *Chosŏn ŭibo* 5 (1935): 39–45. Among a number of calculation methods, Ch'oe Tong relied on the formula (% blood type A + % blood type AB)/(% blood type B + % blood type AB) for the racial index.

101. See, for instance, note 75, my analysis of Arase, "Chōsenjin no taishitsu jinruigakuteki kenkyū," *Chōsen igakkai zasshi* 24, no. 1 (1934): 60–110 (part 1), 111–53 (part 2).

102. Lo, *Doctors within Borders*, 1–18, 104–35. Regarding identity politics under colonialism in East Asia, see Barlow, *Formations of Colonial Modernity in East Asia*.

103. Most of this subsection is based on Paek In-je paksa chŏn'gi kanhaeng wiwŏnhoe, *Sŏngakcha Paek In-je*; Satō Gōzō, *Chōsen yiikushi*; Yi Kyu-sik, "Yu Il-chun ŭi saengae wa hangmun"; and Pak Hyŏng-u, "Taeŭi Kim P'il-sun." It is not an exaggeration to say that Paek In-je has been regarded as one of the most influential doctors of biomedicine in colonial Korea. His legacy continues to be commemorated because of the impact he and his family had on the establishment of modern hospitals and medical schools in Korea. The Paek Hospital (Paek Pyŏngwŏn) was first founded in Seoul in 1932 and expanded to establish college-affiliated hospitals located in four different regions in South Korea, with a total of six thousand beds in 1999. Inje College of Medicine (Inje ŭigwa taehak), which was first established in 1979, was promoted to Inje University (Inje taehakkyo) in 1988 and admitted approximately 2,500 students in 2004. All of these medical and educational accomplishments began with the foundation Paek In-je established in 1946.

104. Paek In-je paksa chŏn'gi kanhaeng wiwŏnhoe, *Sŏngakcha Paek In-je*, 13–35, 69–77.

105. For more about the medical education at the school, see Ki Ch'ang-dŏk, "Ŭihak kyoyuk ŭi hyŏndaehwa kwajŏng."

106. During the first ten years of the Professional School of Medicine, the only Korean instructor was one who briefly taught anatomy. Paek In-je paksa chŏn'gi kanhaeng wiwŏnhoe, *Sŏngakcha Paek In-je*, 52–57, 61, 64–65.

107. Ibid., 66–67.

108. Ibid., 238–40.

109. Ibid., 264–74.

Chapter 4. Lifesaving Water

1. *Maeil sinbo*, September 1, 1911.

2. For the Indan's success in Japan, see Burns, "Marketing Health and Modern Body," 194–98. Sherman Cochran addresses Indan's success in China; Cochran, *Chinese Medicine Men*, 45–47. Patent medicines, or "sold medicine," technically involve some kind of institutional legitimation. Hereafter, I define "patent medicine" as medicines registered with their own brand names, even though the ingredients and efficacy are similar to other ready-made or precompounded medicines. The Japanese government officially controlled patent medicines with detailed regulations beginning in 1912, although the patenting of medicines had been practiced since 1907.

3. *Maeil sinbo*, September 1, 1911. The original caption had the Japanese name of Humane Elixir, Jintan.

4. The Japanese Government-General promulgated a regulation prohibiting "unapplied ads" and "excess and deceptive advertising" in 1912. However, remaining records testify to loose control, particularly for medical ads. Sin In-sŏp and Sŏ Pŏm-sŏk, *Han'guk kwanggosa*, 94.

5. *Maeil sinbo*, January 1, 1916; January 1, 1915; July 21, 1914; October 6, 1915; March 8, 1916; March 31, 1916; November 22, 1916; August 1, 1934.

6. Han'guk kwanggo yŏn'guwŏn, *Han'guk kwanggo paengnyŏn*, 50; Sin In-sŏp and Sŏ Pŏm-sŏk, *Han'guk kwanggosa*, 109–11.

7. Hong Hyŏn-o, *Han'guk yagŏpsa*, 160.

8. Sin In-sŏp and Sŏ Pŏm-sŏk, *Han'guk kwanggosa*, 72. The Meiji government initiated and revised regulations about patent medicine's ingredients, tax, and licensing fees for manufacturers and retailers. See Burns, "Japanese Patent Drug Trade," 6.

9. Newspapers in Japanese, published and owned by the Japanese in Korea, flourished with the growth of Japanese colonial governance. According to statistical reports of the government-general in 1913, of the twenty-five newspapers circulated in Korea, most had been launched between 1900 and 1909. Han'guk kwanggo yŏn'guwŏn, *Han'guk kwanggo paengyŏn*, 65. The daily circulation of all the Japanese newspapers in Korea reached thirty-four thousand copies a day in 1909. Sin In-sŏp and Sŏ Pŏm-sŏk, *Han'guk kwanggosa*, 67.

10. *Hansŏng chubo* was published between 1886 and 1888.

11. Sin In-sŏp, *Han'guk kwanggo paltalsa*, 18–19.

12. Sin In-sŏp and Sŏ Pŏm-sŏk, *Han'guk kwanggosa*, 120. This number includes 97,620 domestic and 17,751 overseas copies.

13. For journal circulation, see ibid., 120–21.

14. Haynes, "Creating the Consumer?"

15. Cochran, *Chinese Medicine Men*, 11–13.

16. Oh, "Consuming the Modern," 11.

17. Ibid., 20.

18. Pak Yun-jae, *Han'guk kŭndae ŭihak ŭi kiwŏn*, 322–30; Burns, "Japanese Patent Drug Trade," 8–10.

19. The Korean government, then, authorized only two professions—pharmacist (*yakchesa*) and drug seller (*yakchongsang*)—in addition to physician. Pak Yun-jae, *Han'guk kŭndae ŭihak ŭi kiwŏn*, 139.

20. Burns, "Japanese Patent Drug Trade," 9.

21. Pak Yun-jae, *Han'guk kŭndae ŭihak ŭi kiwŏn*, 57–68.

22. Ibid., 247–67.

23. Hong Hyŏn-o, *Han'guk yagŏpsa*, 7–15.

24. Ibid., 160–61.

25. Ibid., 191–94.

26. Ibid., 5–8.

27. These periodicals were published in Japanese. Ibid., 184–85.

28. Nihon denpō tsūshinsha (Japan Telegraphic Communication Company), also known as Dentsu, first established its Korean branch on April 30, 1906, then nearly monopolized the market of advertising in Korea. Sin In-sŏp and Sŏ Pŏm-sŏk, *Han'guk kwanggosa*, 176. A few more Japanese advertisement agencies opened during 1920–45. Ibid., 132.

29. Ibid., 192–93.

30. Ibid., 179–85.

31. Ibid., 193.

32. For the history of Lifesaving Water, see Ye Chong-sŏk, *Hwalmyŏngsu 100 nyŏn sŏngjang* and Hong Hyŏn-o, *Han'guk yagŏpsa*, 20–22.

33. Hong Hyŏn-o, *Han'guk yagŏpsa*, 3–4; Ye Chong-sŏk, *Hwalmyŏngsu 100 nyŏn sŏngjang*, 16–17.

34. Ye Chong-sŏk, *Hwalmyŏngsu 100 nyŏn sŏngjang*, 19.

35. Ibid., 35. In 1915, ¥1 was roughly the equivalent of US$1, and 1/100 of ¥1 was 1 *chŏn* (*sen* in Japanese). In the early 1920s, 10 kg of rice cost ¥2.5. In 1915, Seoul was a city with a population of 600,000, including 100,000 Japanese. The entire population of Korea was 15,930,000, including 291,000 Japanese in 1914. Hong Hyŏn-o, *Han'guk yagŏpsa*, 161; Sin In-sŏp and Sŏ Pŏm-sŏk, *Han'guk kwanggosa*, 73.

36. Ye Chong-sŏk, *Hwalmyŏngsu 100 nyŏn sŏngjang*, 31.

37. Ibid., 22.

38. See a series of ads in *Maeil sinbo*, for instance, February 4, 1910; July 2, 1912; June 1, 1918; and November 17, 1918.

39. Ye Chong-sŏk, *Hwalmyŏngsu 100 nyŏn sŏngjang*, 62–63.

40. *Chosŏn ilbo*, August 11, 1936.

41. Han'guk kwanggo yŏn'guwŏn, *Han'guk kwanggo paengyŏn*, 270.

42. See, for instance, *Maeil sinbo*, June 17, 1913.

43. One *chŏn* was the equivalent of one US cent at that time.

44. Hong Hyŏn-o, *Han'guk yagŏpsa*, 12–13. For the Japanese Residency-General's control of patent medicines, including the Elixir for Clearing the Heart and Guarding Life, see Pak Yun-jae, "Ch'ŏngsim pomyŏngdan nonjaeng."

45. Korea Press Foundation provides a digital archive of a series of newspapers published before 1945 at http://www.bigkinds.or.kr/mediagaon/goNewsDirectory .do. I searched the elixir's ads in *Taehan maeil sinbo*.

46. *Taehan maeil sinbo*, November 14, 1909.

47. *Maeil sinbo*, April 28, 1911.

48. *Taehan maeil sinbo*, November 14, 1909.

49. For the endemic nature of pulmonary distomatosis, see Pak Yun-jae, *Han'guk kŭndae ŭihak ŭi kiwŏn*, 245–46.

50. *Taehan maeil sinbo*, November 14, 1909.

51. The scale of the medical market in Manchuria is not entirely known. According to Cho Hyŏn-bŏm and Chin Chŏn-gyu, in Wangqing County in northeast China, both Chinese and Korean merchants divided the market equally in the 1930s. Among the population of two hundred thousand, Korean immigrants represented approximately 40 percent. By the time of Manchukuo's establishment in 1932, over one hundred apothecaries, eighty-nine clinics of traditional medicine, and itinerant practitioners provided consultation on the health concerns of residents. Cho Hyŏn-bŏm and Chin Chŏn-gyu, "Wangch'ŏnghyŏn ŭi kaein chungyakpang."

52. *Mansŏn ilbo* was first published in 1937, combining two previous local newspapers, *Kando ilbo* (Daily News of Kando), which appeared in the 1920s, and *Manmong ilbo* (Daily News of Manchuria and Mongolia), first printed in 1933 and published in Korean. *Mansŏn ilbo* is one of the most significant sources of information on the daily lives of Korean immigrants in Manchuria during the first half of the twentieth century. Extant copies are digitalized and accessible via the National Library of Korea.

53. For instance, *Mansŏn ilbo*, December 5, 1939; January 7, 13, 28, 1940.

54. *Mansŏn ilbo*, January 3, 1940.

55. *Mansŏn ilbo*, December 5, 1939.

56. "Kŏnjae yich'ul pangji k'o Chosŏn nae sugŭp t'ongje," *Mansŏn ilbo*, December 15, 1939.

57. *Mansŏn ilbo*, January 13, 1940.

58. *Mansŏn ilbo*, December 7, 1938; January 13, 1939.

59. "Nambungman Chosŏn kaech'ŏngmin chiptan purak tapsagi," *Mansŏn ilbo*, January 4, 1940.

60. *Mansŏn ilbo*, January 7, 1940.

61. *Mansŏn ilbo*, January 1, 1940.

62. *Mansŏn ilbo*, December 13, 1939.

63. Beginning on January 1, 1940, a series of interviews was published in *Mansŏn ilbo*. Another example is December 1 and 8, 1940.

64. Kang Kyŏng-ae, "Sogŭm." For an introduction to Kang Kyŏng-ae's literary works, see Hughes et al., *Rat Fire*, 212–14.

65. Hong Hyŏn-o, *Han'guk yagŏpsa*, 15–16.

66. Ibid., 112–13.

67. Ibid.

68. Ibid., 47–48.

69. Ibid., 114.

70. Ye Chong-sŏk, *Hwalmyŏngsu 100 nyŏn sŏngjang*, 38.

71. Yang Chŏng-p'il, "Hanmal ilche ch'o kŭndaejŏk yagŏp," 205–6.

72. For detailed stories about Yu Ir-han and the Yuhan Company, see Cho Sŏng-gi, *Yu Ir-han p'yŏngjŏn*; Yu Ir-han Chŏn'gi P'yŏnjip Wiwŏnhoe, *Nara sarang ŭi ch'am kiŏbin*; and Kim Kyo-sik, *Yu Ir-han, Yuhan Yanghaeng gŭrup*.

73. Cho Sŏng-gi, *Yu Ir-han p'yŏngjŏn*, 158–66.

74. Ibid., 170–77, and see Yu Ir-han Chŏn'gi P'yŏnjip Wiwŏnhoe, *Nara sarang ŭi ch'am kiŏbin*, 181.

75. For the economic transition in the late 1910s and early 1920s, see Eckert, *Offspring of Empire*, 27–59. For the quote, see 40.

76. Ibid., 42.

77. Yu Ir-han Chŏn'gi P'yŏnjip Wiwŏnhoe, *Nara sarang ŭi ch'am kiŏbin*, 184.

78. Hong Hyŏn-o, *Han'guk yagŏpsa*, 53–55.

79. Ibid., 57.

80. Sin In-sŏp and Sŏ Pŏm-sŏk, *Han'guk kwanggosa*, 197.

81. Sin In-sŏp, "Han'guk ŭi kukche kwanggosa."

82. Sin In-sŏp and Sŏ Pŏm-sŏk, *Han'guk kwanggosa*, 194. Interestingly, Kang Han-in's exceptional experience at the Yuhan Company helped him join the administration of the first South Korean president, Syngman Rhee. Kang Han-in served as Syngman Rhee's press secretary.

83. *Tonga ilbo*, July 9, 1928. The licensed pharmacist Na Ch'ansu, who had worked at the Government-General Hospital, was also mentioned.

84. Hong Hyŏn-o, *Han'guk yagŏpsa*, 58. Yuhan's relationship with Abbott enabled the firm to reach out to Taiwan, China, Manchuria, and Japan in the late 1930s.

85. Yu Ir-han Chŏn'gi P'yŏnjip Wiwŏnhoe, *Nara sarang ŭi ch'am kiŏbin*, 193–94, 215.

86. Hong Hyŏn-o, *Han'guk yagŏpsa*, 67–71.

87. Advertisements for Satsutarimu frequently appeared in *Mansŏn ilbo* between 1940 and 1943. For a case study on the Japanese pharmaceutical company and Japanese research on Salvarsan, see Banyu Pharmaceutical, *Medicine for the People*.

88. Hong Hyŏn-o, *Han'guk yagŏpsa*, 67–68.

89. *Mansŏn ilbo*, December 3, 1940.

90. Yu Ir-han Chŏn'gi P'yŏnjip Wiwŏnhoe, *Nara sarang ŭi ch'am kiŏbin*, 518.

91. Ibid., 519–20.

92. Ibid., 190; Cho Sŏng-gi, *Yu Ir-han p'yŏngjŏn*, 216.

93. Yu Ir-han Chŏn'gi P'yŏnjip Wiwŏnhoe, *Nara sarang ŭi ch'am kiŏbin*, 201.

94. Ibid., 207–8. Interestingly, Seoul and North Korean residents were most of the domestic shareholders.

95. Kim Kyo-sik, *Yu Ir-han, Yuhan Yanghaeng gŭrup*, 148–49.

96. Yu Ir-han Chŏn'gi P'yŏnjip Wiwŏnhoe, *Nara sarang ŭi ch'am kiŏbin*, 222.

97. Eckert, *Offspring of Empire*, 58.

Chapter 5. Fire Illness, or Hwabyŏng

1. Min Sŏng-gil and Kim Chin-hak, "Pogilto esŏ ŭi hwabyŏng," 459 in particular.

2. Yi Si-hyŏng, "Hwabyŏng ŭi kaenyŏm e taehan yŏn'gu."

3. Min Sŏng-gil et al., "Hwatpyŏng e taehan chindanjŏk yŏn'gu," 654.

4. Here I refer to the second (1968) and the third (1980) editions.

5. Li Jianmin, "'Bencao gangmu. huobu' kaoshi."

6. Furth, "Physician as Philosopher."

7. Min Sŏng-gil, *Hwabyŏng yŏn'gu*, 18–19.

8. Sin Yŏng-gŭn, "Kim Yŏng-hwan, taesŏn ch'ulma sŏnŏn," *Yŏnhap nyusŭ*, July 5, 2012.

9. Kim Yŏl-gyu, *Han'gugin ŭi hwa*.

10. Kim T'ae-yŏl, "Yŏsŏng amhwanja 85 % sŭt'ŭresŭ ro hwabyŏng kkaji," *Chosŏn ilbo*, March 23, 2011.

11. For instance, "Mun Ik-hwan, puksŏ p'ŭrakch'i ro mora hwabyŏng sumjyŏ," *Donga ilbo*, September 24, 2011; "Angibu p'ŭrakch'i ro mon pukhan e pun'gae," *JoongAng Sunday*, September 18, 2011.

12. "Constrained fire" is romanized as *urhwa* in Korean or *yuhuo* in Chinese.

13. Ch'oe Po-sik, "Pak Kyŏng-ni ŭi ttal, Kim Chi-ha ŭi anae," *Chosŏn ilbo*, February 28, 2011.

14. *Chŏngjo sillok*, 1800. 6. 16.

15. For more details about *hwabyŏng* in Korean court records, see Kim Chong-u, Hyŏn Kyŏng-ch'ŏl, and Hwang Ŭi-wan, "Hwabyŏng ŭi kiwŏn e kwanhan koch'al."

16. Porter, "'Expressing Yourself Ill,'" 277.

17. Watters, *Crazy Like Us*.

18. See T'onggyech'ŏng, "Samang wŏnin t'onggye," 2003–2014.

19. For a bibliography and a brief introduction to past research, see Min Sŏng-gil, *Hwabyŏng yŏn'gu*, and Min, "Hwa-Byung: Anger Syndrome," especially 3–4.

20. Pang, "Hwabyung"; Lin et al., *"Hwa-byung."*

21. Yi Si-hyŏng is a well-known psychiatrist, popular author, and now a director of a healing center in South Korea. He was affiliated with one of the two most privileged private universities in South Korea and has played a role in popularizing *hwabyŏng* since the 1970s.

22. Yi Si-hyŏng, "Hwabyŏng ŭi kaenyŏm e taehan yŏn'gu."

23. Min Sŏng-gil is one of the most well-known specialists of *hwabyŏng*. Since the 1980s, Min Sŏng-gil has published dozens of articles and a book about *hwabyŏng*, calling for Korean psychiatrists' attention to this seemingly indigenous illness.

24. Min Sŏng-gil et al., " Hwatpyŏng e taehan chindanjŏk yŏn'gu," 653–55.

25. Ibid., 655.

26. Ibid., 661.

27. Min Sŏng-gil and Kim Chin-hak, "Pogilto esŏ ŭi hwabyŏng e taehan yŏn'gu."

28. Ibid., 466.

29. Min Sŏng-gil, "Hwatpyŏng kwa han," 1194.

30. Min Sŏng-gil and Kim Chin-hak, "Pogilto esŏ ŭi hwabyŏng e taehan yŏn'gu," 464.

31. Min Sŏng-gil et al., " Hwatpyŏng e taehan chindanjŏk yŏn'gu," 653.

32. Min Sŏng-gil, "Hwatpyŏng kwa han," 1192.

33. Ibid., 1198.

34. Ibid., 1195.

35. For instance, see Kim Chong-u et al., "Hwabyŏng kwa hwabyŏng chuyo uulchŭng." Later research about *hwabyŏng* published in *Tongŭi singyŏng chŏngsin kwahak hoeji*, *Taehan ŭihak hyŏphoeji* (Journal of Korean Medical Association), and *Singyŏng chŏngsin ŭihak* (Journal of Korean Neuropsychiatric Association) often relies on Min Sŏng-gil for the definition of *hwabyŏng*.

36. Ch'ŏn I-du, *Han ŭi kujo yŏn'gu*.

37. For instance, Son, *Haan of Minjung Theology*, and Hee-soo Kim, "Roots of Han and Its Healing."

38. Willoughby, "Sound of Han."

39. Ch'oe Kil-sŏng, *Han'gugin ŭi han*.

40. Ko Ŭn, "Han ŭi kŭkpok ŭl wihayŏ," 33.

41. Min Sŏng-gil, *Hwabyŏng yŏn'gu*, 219.

42. Lin, " Hwa-byung: A Korean Culture-Bound Syndrome?"

43. Min Sŏng-gil, *Hwabyŏng yŏn'gu*, 217.

44. Ibid., appendix, 219–22.

45. Ibid. See also Min Sŏng-gil et al., "Hwabyŏng ch'ŏkto."

46. Min Sŏng-gil, *Hwabyŏng yŏn'gu*, 176.

47. Ibid., 77.

48. Ibid., 83.

49. Ibid., 186–87.

50. Min Sŏng-gil, So Ŭn-hŭi, and Pyŏn Yong-uk, "Chŏngsin'gwa ŭisa wa hanŭisa."

51. Kim Ch'un-ho, "Chŏn'guk hanŭigwa taehak chaejik kyosu," *Minjok ŭihak sinmun* (Minjok medicine news), May 14, 2015.

52. Song Kyŏng-sŏp, "Hanŭihak ŭi palchŏn," 28–30.

53. Kim Chong-u et al., "Hwabyŏng kwa hwabyŏng chuyo uulchŭng," 6–7.

54. Regarding the specialization of traditional medicine in Korea, see, for instance, Han Ch'ang-ho, "Hanŭisa chŏnmunŭi chedo sihaeng 10 nyŏn," *Minjok ŭihak sinmun*, February 16, 2012.

55. "Uulchŭng yak sobi 5 nyŏnsae 52% nŭrŏtta," *Yŏnhap nyusŭ*, July 26, 2009.

56. http://www.theksad.or.kr/network/about.php?q=about01. The name of the society was changed to the Korean Soceity for Affective Disorders.

57. *Ul* is pronounced *yu* in Chinese.

58. For more discussion on constraint (*ul*), see the special issue and the introduction by Scheid, "Constraint as a Window."

59. Kim Chong-u, *Hwabyŏng ŭro put'ŏ ŭi haebang*, 109–10.

60. Kim Chong-u has played a leading role in establishing *hwabyŏng* studies in Korea for the past twenty years. See, for instance, http://www.hwabyung.kr.

61. Kim Chong-u and Hwang Ŭi-wan, "Hanŭihak esŏ pon hwabyŏng," 9–10.

62. Ibid., 11–13.

63. Ibid., 11.

64. Kim Chong-u, *Hwabyŏng ŭro put'ŏ ŭi haebang*, 7.

65. Yim Chae-hwan, Kim Chong-u, and Hwang Ŭi-wan, "Hanŭihakchŏk hwatpyŏng ch'iryo."

66. Holmes and Rahe, "Social Readjustment Rating Scale"; Camille Lloyd, "Life Events and Depressive Disorder." See also Yim Chae-hwan, Kim Chong-u, and Hwang Ŭi-wan, "Hanŭihakchŏk hwatpyŏng ch'iryo," 53–54.

67. Yi Hŭi-yŏng et al., "Hwabyŏng ŭi chindan mit pyŏngjŭng yuhyŏng."

68. Ibid., 12.

69. Karchmer, "Chinese Medicine in Action," 226–30.

70. For instance, at http://www.omstandard.com. This web page no longer exists, but for the details about the OMS-prime as a diagnostic standard of *hwabyŏng*, see Kim Chong-u, *Hwabyŏng ŭro put'ŏ ŭi haebang*, 246–54, and Ch'oe Sŭng-hun, "Int'ŏnet kiban hanŭi chindan chŏnmunga sisŭt'em kaebal."

71. Kim Chong-u et al., "Hwabyŏng kwa hwabyŏng chuyo uulchŭng."

72. Ibid., 12.

73. Traditional medicine's effort to establish an objective and standardized guide for diagnosis continues after the completion of HBDIS. See Kim Chong-u et al., "Hwabyŏng imsang chillyo chich'im kaebal yŏn'gu (1)"; Kim Chong-u et al., " Hwabyŏng yŏkhak yŏn'gu charyo"; Chŏng Myŏng-hŭi et al., "Hwabyŏng hanŭi p'yŏngga togu kaebal."

74. Chŏng Sŏn-yong et al., "Hwangnyŏn haedokt'ang," 2.

75. Kim Chong-u and Hwang Ŭi-wan, "Hanŭihak esŏ pon hwabyŏng ŭi haesŏk."

76. Min, "Hwa-Byung: Anger Syndrome"; Min, "Differences in Temperament."

77. Hwang Ŭi-wan, *Hwabyŏng kŭkpok p'ŭrojekt'ŭ*.

78. Ibid., 11.

79. See Kim Chong-u and Hwang Ŭi-wan, "Hanŭihak esŏ pon hwabyŏng ŭi haesŏk," and Kim Chong-u and Hwang Ŭi-wan, "Hwabyŏng hwanja ŭi hanŭihakchŏk ch'iryo."

80. Ch'oe Yŏng-jin et al., *Manbyŏng ŭi kŭnwŏn hwabyŏng*, 28.

81. Yun Yong-sŏp, *Algo namyŏn swiun hwabyŏng haesŏl*.

82. Kang Yong-wŏn, *Annyŏng, uulchŭng*, 252.

83. Hwang Ŭi-wan, *Hwabyŏng kŭkpok p'ŭrojekt'ŭ*, 11; Ch'oe Yŏng-jin et al., *Manbyŏng ŭi kŭnwŏn hwabyŏng*, 5.

84. Kang Yong-wŏn, *Annyŏng, uulchŭng*, 254.

85. See, for instance, Min Sŏng-gil, "Hwatpyŏng kwa han," 1195.

86. Hwang Ŭi-wan, *Hwabyŏng kŭkpok p'ŭrojekt'ŭ*, appendix 1.

87. Yim Chae-hwan, Kim Chong-u, and Hwang Ŭi-wan, "Hanŭihakchŏk hwatpyŏng ch'iryo."

88. Hwang Ŭi-wan, *Hwabyŏng kŭkpok p'ŭrojekt'ŭ*, 14–56.

89. Ibid., appendix 4.

90. Yun Yong-sŏp, *Algo namyŏn swiun hwabyŏng haesŏl*, 171–239.

91. Kang Yong-wŏn, *Annyŏng, uulchŭng*, 175.

92. Ibid., 189.

93. Ibid., 189–99.

94. Kleinman, *Illness Narratives*.

95. Charon, *Narrative Medicine*. For more discussions about narrative medicine, see Mayers, "Commentary on Myra's Life"; Reichert et al., "Narrative Medicine"; DasGupta, "Art of Medicine"; and Charon and Wyer, "Art of Medicine."

96. Sandy Banks, "Damage Went Deep," *Los Angeles Times*, May 1, 2012.

97. Porter, "'Expressing Yourself Ill,'" 277.

98. Ibid., 291.

99. Ibid., 284.

100. Yi Kyŏng-nim, "Korean Women," 96. As a well-established poet in South Korea, Yi Kyŏng-nim made her debut in 1989 and is the author of five volumes of poetry, one volume of essays, and a novel. A volume of poetry is available in English. Yi Kyŏng-nim, *New Season Approaching*.

Epilogue

1. "Airs, Waters, Places," in G. E. R. Lloyd, *Hippocratic Writings*, 148.

2. Kuriyama, *Expressiveness of the Body*, 272.

3. "Airs, Waters, Places," in G. E. R. Lloyd, *Hippocratic Writings*, 148.

4. Charon, *Narrative Medicine*, 9.

5. Pak Sun-yŏng, "Ilche singmin t'ongch'i ha ŭi," 201.

6. For terms like "separation" and "purity," I referred to Bruno Latour's criticism of the modern's dualism. Latour, *We Have Never Been Modern*, 46–48, 127–29.

7. Yi Sŏ-u's poem is included in Chŏng Yag-yong, "Hubak, modan" 厚朴 牡丹, in *Aŏn kakpi*. See chapter 1 for more discussion.

8. Yi Kyŏng-nim, "Korean Women," 96.

Bibliography

Primary Sources

Books

Classical Chinese

Chŏng Yag-yong 丁若鏞. *Aŏn kakpi* 雅言覺非 (Realizing faults by rectifying words). 1819. Annotated and translated into modern Korean by Chŏng Hae-ryŏm 정해렴. Seoul: Hyŏndae sirhaksa, 2005.

———. *Yŏyudang chŏnsŏ* 與猶堂全書 (The collected works of Yŏyudang). 1934–1938. http://www.krpia.co.kr.

Chosŏn wangjo sillok 朝鮮王朝實錄 (Veritable records of the Chosŏn dynasty). http://sillok.history.go.kr.

Han'guk kwahak kisulsa charyo taegye: ŭiyakhakp'yŏn 韓國科學技術史資料大系: 醫藥學篇 (The grand series of the history of Korean science and technology: Medicine). 50 vols. Seoul: Yŏgang ch'ulp'ansa, 1988.

Hŏ Chun 許浚. *Tongŭi pogam* 東醫寶鑑 (Precious mirror of Eastern medicine). 1613. Seoul: Yŏgang ch'ulp'ansa, 2005.

Hwang To-yŏn 黃度淵. *Pangyak happ'yŏn* 方藥合編 (Compendium of prescriptions). 1885. In *The Grand Series of the History of Korean Science and Technology: Medicine*, vol. 27.

———. *Ŭijong sonik* 醫宗損益 (Foundations of medicine revised). 1868. In *The Grand Series of the History of Korean Science and Technology: Medicine*, vols. 25–26.

Hyangyak chipsŏngbang 鄉藥集成方 (Standard prescriptions of local botanicals). 1433. In *The Grand Series of the History of Korean Science and Technology: Medicine*, vol. 3.

Hyangyak kugŭppang 鄉藥救急方 (Prescriptions of local botanicals for emergency use). Ca. 1236. In *The Grand Series of the History of Korean Science and Technology: Medicine*, vol. 1.

Kang Myŏng-gil 康命吉. *Chejung sinp'yŏn* 濟眾新編 (New compilation for benefiting people). 1799. In *The Grand Series of the History of Korean Science and Technology: Medicine*, vol. 18.

Koryŏsa 高麗史 (The history of Koryŏ). Ca. fifteenth century. http://www.krpia.co.kr.

Sŏng Chu-bong 成周鳳. *Hanbang ŭihak kangsŭpsŏ* 漢方醫學講習書 (A textbook of traditional medicine). Taejŏn, South Korea: T'aep'yŏngdang, 1936.

Wŏn Chi-sang 元持常. *Tongŭi sasang sinp'yŏn* 東醫四象新編 (Newly edited Four Constitutions of Eastern medicine). Kyŏngsŏng: Munusa, 1929. http://www.krpia.co.kr.

Yi Che-ma 李濟馬. *Tongŭi suse powŏn* 東醫壽世保元 (Longevity and life preservation in Eastern medicine). 1894. Seoul: Yŏgang ch'ulp'ansa, 1992. Translated into English by Ch'oe Sŭng-hun. Seoul: Kyŏnghŭi University Press, 1996.

Zhang Ji 張機. *Shanghanlun: On Cold Damage*. Ca. 196–220. Translated into English by Craig Mitchell, Feng Ye, and Nigel Wiseman. Brookline, MA: Paradigm, 1999.

Korean

Cho Hŏn-yŏng 조헌영, Chang Ki-mu 장기무, Chŏng Kŭn-yang 정근양, Yi Ŭr-ho 이을호, Sin Kil-gu 신길구, and Kang P'il-mo 강필모. *Hanŭihak ŭi pip'an kwa haesŏl* 한의학의비판과해설 (Criticism and explanation about traditional medicine). Ca. 1936. Seoul: Sonamu, 1997.

JOURNALS AND NEWSPAPERS

Japanese

Chōsen igakkai zasshi 朝鮮醫學會雜誌 (Journal of the Chōsen Medical Association), 1911–42.

Mansen no ikai 滿鮮之醫界 (Medical world of Manchuria and Korea), 1926–40.

Korean

Chosŏn ilbo 조선일보 (Korean Daily), 1920–40.
Chosŏn ŭibo 조선의보 (Korean medical journal), 1931–37.
Chosŏn ŭihakkye 조선의학계 (Association of Korean medicine), 1918–19.

Hanbang ŭiyak 한방의약 (Traditional medicine), 1937–42.
Hanbang ŭiyakkye 한방의약계 (Association of Korean traditional medicine), 1913–14.

Maeil sinbo 매일신보 (Daily News), 1911–45.
Mansŏn ilbo 만선일보 (Daily News of Manchuria and Korea), 1939–41.

Taehan maeil sinbo 대한매일신보 (Korean Daily News), 1905–10.

Tonga ilbo 동아일보 (East Asian Daily), 1921–40.

Tongsŏ ŭihak yŏn'guhoe wŏlbo 동서의학연구회월보 (Monthly report of the Association for Studying Eastern and Western Medicine), 1924.

Secondary Sources

American Psychiatric Association. *Diagnostic and Statistical Manual of Mental Disorders.* 2nd ed. Washington, DC: American Psychiatric Association, 1968 (3rd ed. 1980).

An Sang-u 안상우. "Koryŏ ŭisŏ 'Piye paegyobang' ŭi kojŭng" 고려의서비예백요방의 고증 (The identification of *Biyebackyobang*, an ancient Korean medical book of the Koryo period). *Sŏjihak yŏn'gu* 22 (2001): 325–50.

———. "Ŭich'ang ŭro parabon chŏnt'ong kwa kŭndae ŭi twitkolmok" 의창으로바라 본전통과근대의뒷골목 (The backstreet of tradition and modernity viewed from a window of medicine). *Kirogin* 26 (2014): 62–67.

An Sang-u 안상우 and Kim Hyŏn-gu 김현구. "Sasang ŭihak ŭi imsang ŭngyong kwa chŏbyŏn hwaktae: Wŏn Chi-sang ŭi 'Tongŭi sasang sinp'yŏn' ŭl chungsimŭro" 사 상의학의임상응용과저변확대:원지상의"동의사상신편"을중심으로 (Clinical application of Sasang constitutional medicine and the spread of its use: Focusing on *Dongui sasang sinpyeon*). *Han'guk ŭisahak hoeji* 25, no. 2 (2012): 97–103.

Atkins, E. Taylor. *Primitive Selves: Koreana in the Japanese Colonized Gaze, 1910–1945.* Berkeley: University of California Press, 2010.

Banyu Pharmaceutical. *Medicine for the People: The First 85 Years of Banyu Pharmaceutical, 1915–2000.* Tokyo: Banyu Pharmaceutical, 2001.

Barlow, Tani, ed. *Formations of Colonial Modernity in East Asia.* Durham, NC: Duke University Press, 1997.

Barnes, Linda L. *Needles, Herbs, Gods, and Ghosts: China Healing and the West to 1848.* Cambridge, MA: Harvard University Press, 2007.

Burns, Susan L. "The Japanese Patent Drug Trade in East Asia: Consumer Culture, Colonial Medicine, and Imperial Modernity." Paper presented at "Medicine, Politics, and Culture in the Japanese Empire," University of Chicago, May 11–12, 2012. This article will be published in *Gender, Health, and History in Modern East Asia*, edited by Angela K. C. Leung and Izumi Nakayama. Hong Kong: Hong Kong University Press, forthcoming.

———. "Marketing Health and the Modern Body: Patent Medicine Advertisements in Meiji-Taishō Japan." In *Looking Modern: East Asian Culture from Treaty Ports to World War II*, edited by Jennifer Purtle and Hans Bjarne Thomsen, 179–202. Chicago: Art Media Resources, 2009.

Chambers, David W., and Richard Gillespie. "Locality in the History of Science: Colonial Science, Technoscience, and Indigenous Knowledge." *Osiris* 15 (2000): 221–40.

Charon, Rita. *Narrative Medicine: Honoring the Stories of Illness.* Oxford: Oxford University Press, 2006.

Charon, Rita, and Peter Wyer. "The Art of Medicine: Narrative Evidence Based Medicine." *Lancet* 371, no. 9609 (2008): 296–97.

Chen, Hsiufen. "Nourishing Life, Cultivation and Material Culture in the Late Ming: Some Thoughts on 'Zunsheng bajian 遵生八牋' (*Eight Discourses on Respecting Life*, 1591)." *Asian Medicine: Tradition and Modernity* 4, no. 1 (2008): 29–45.

Chi Ch'ang-yŏng 지창영. "'Chejung sinp'yŏn' ŭi inyong pangsik e taehan yŏn'gu" 제증신편"의인용방식에대한연구 (A study on citation methods of *Jejoongshinpyeon*). *Han'guk ŭisahak hoeji* 21, no. 1 (2008): 83–88.

Cho Hak-chun 조학준. "Ilche kangjŏmgi ŭi hanŭihak kyojae chung hanain 'Hanbang ŭihak kangsŭpsŏ'" 일제강점기의한의학교재중하나인"한방의학강습서 (A lecture book on traditional Korean medicine in the period of Japanese occupation, *Hanbang eihak gangweupseo*). *Han'guk ŭisahak hoeji* 22, no. 1 (2009): 77–104.

Cho Hŏn-yŏng 조헌영. *T'ongsok hanŭihak wŏllon* 통속한의학원론 (Layman's guide to traditional medicine). Seoul: Hakwŏnsa, 2002.

Cho Hyŏn-bŏm 조현범 and Chin Chŏn-gyu 진전규. "Wangch'ŏnghyŏn ŭi kaein chungyakpang" 왕청현의개인중약방 (The private Chinese apothecaries in Wangqing Prefecture). In *Haebang chŏn yŏnbyŏn kyŏngje* 해방전연변경제 (The economy of Yanbain before the liberation), edited by Yanbian Chaoxianzu Zizhizhou, 156–60. Yanji, China: Yanbian renmin chubanshe, 1994.

Cho Sŏng-gi 조성기. *Yu Ir-han p'yŏngjŏn* 유일한평전 (A critical biography of Yu Ir-han). Seoul: Chagŭn ssiat, 2005.

Ch'oe Kil-sŏng 최길성. *Han'gugin ŭi han* 한국인의한 (The *han* of Koreans). Seoul: Yejin, 1991.

Ch'oe Sŭng-hun 최승훈. "Int'ŏnet kiban hanŭi chindan chŏnmunga sisŭt'em kaebal" 인터넷기반한의진단전문가시스템개발 (Development of web-based diagnosis expert system of traditional Oriental medicine). *Tongŭi saengni pyŏngnihak hoeji* 16, no. 3 (2002): 528–31.

Ch'oe Yŏng-jin 최영진, Yi Hwa-jin 이화진, Kim Chun-hong 김준홍, and Hwang Man-gi 황만기. *Manbyŏng ŭi kŭnwŏn hwabyŏng kwa sŭt'ŭresŭ t'ap'a* 만병의근원화병과스트레스타파 (Breaking down *hwabyŏng* and stress, the origins of all diseases). Seoul: Pukp'ia, 2009.

Chŏn Hye-ri 전혜리. "1934 nyŏn hanŭihak puhŭng nonjaeng" 1934년한의학부흥논쟁 (Modernizing Korean medicine: The debate on the revival of Korean medicine in colonial Korea in the 1930s). *Han'guk kwahaksahak hoeji* 33, no. 1 (2011): 41–89.

Ch'ŏn I-du 천이두. *Han ŭi kujo yŏn'gu* 한의구조연구 (A study of the structure of *han*). Seoul: Munhak kwa chisŏngsa, 1993.

Chŏng Chi-hun 정지훈. "Hanŭi haksul chapchi rŭl chungsimŭro salp'yŏ pon ilche sidae hanŭihak ŭi haksulchŏk kyŏnghyang" 한의학술잡지를중심으로살펴본일제시대한의학의학술적경향 (Research into academic journals of Oriental medicine in the era of Japanese imperialism). *Han'guk ŭisahak hoeji* 17, no. 1 (2004): 195–253.

Chŏng Myŏng-hŭi 정명희, Yi Sang-nyong 이상룡, Kang Wi-ch'ang 강위창, and Chŏng In-ch'ŏl 정인철. "Hwabyŏng hanŭi p'yŏngga togu kaebal ŭl wihan kich'o yŏn'gu" 화병한의평가도구개발을위한기초연구 (Preliminary study to develop the instrument of Oriental medical evaluation for Hwa-byung). *Tongŭi sin'gyŏng chŏngsin kwahak hoeji* 21, no. 2 (2010): 141–55.

Chŏng Sŏn-yong 정선용, Kim Chong-u 김종우, Yi Chŏng-nyun 이정륜, Chang Hyŏn-ho 장현호, Kim Hyŏn-t'aek 김현택, and Hwang Ŭi-wan 황의완. "Hwangnyŏn haedokt'ang i uulchŭng mohyŏng tongmul ŭi uul sŏnghyang mit PVN ŭi c-Fos palhyŏn e mich'inŭn hyogwa" 황련해독탕이우울증모형동물의우울성향및 PVN 의 c-Fos 발현에미치는효과 (Effects of Hwangryeon haedokt'ang on depression and c-Fos expression in paraventricular nucleus of the brain in the chronic mild stress treated rats). *Tongŭi sin'gyŏng chŏngsin kwahak hoeji* 14, no. 1 (2003): 1–16.

Chŏn'guk hanŭigwa taehak sasang ŭihak kyosil 전국한의과대학사상의학교실. *Sasang ŭihak* 사상의학 (Four Constitutions medicine). Seoul: Chipmundang, 2008.

Clark, Donald N. "Sino-Korean Tributary Relations under the Ming." In *The Cambridge History of China*, edited by Denis Twitchett and Frederick W. Mote, vol. 8, 272–300. New York: Cambridge University Press, 1998.

Cochran, Sherman. *Chinese Medicine Men: Consumer Culture in China and Southeast Asia*. Cambridge, MA: Harvard University Press, 2006.

Contemporary Chinese Dictionary. Beijing: Foreign Language Teaching and Research Press, 2002.

Cooper, Alix. *Inventing the Indigenous: Local Knowledge and Natural History in Early Modern Europe*. New York: Cambridge University Press, 2007.

———. "Latin Words, Vernacular Worlds: Language, Nature, and the 'Indigenous' in Early Modern Europe." *East Asian Science, Technology, and Medicine* 26 (2007): 17–39.

Croizier, Ralph. "The Ideology of Medical Revivalism in Modern China." In *Asian Medical Systems, a Comparative Study*, edited by Charles Leslie, 341–55. Berkeley: University of California Press, 1976.

Crossgrove, William. "The Vernacularization of Science, Medicine, and Technology in Late Medieval Europe: Broadening Our Perspectives." *Early Science and Medicine* 5, no. 1 (2000): 47–63.

Daidoji, Keiko. "Treating Emotion-Related Disorders in Japanese Traditional Medicine: Language, Patients, and Doctors." *Culture, Medicine, and Psychiatry* 37 (2013): 59–80.

DasGupta, Sayantani. "The Art of Medicine: Narrative Humility." *Lancet* 371, no. 9617 (2008): 980–81.

Eckert, Carter J. *Offspring of Empire: The Koch'ang Kims and the Colonial Origins of Korean Capitalism, 1876–1945*. Seattle: University of Washington Press, 1997.

Epler, Dean C., Jr. "The Concept of Disease in an Ancient Chinese Medical Text, *The Discourse on Cold-Damage Disorders (Shang-han Lun)*." *Journal of the History of Medicine and the Allied Sciences* 43, no. 1 (1988): 8–35.

Fan, Fa-ti. *British Naturalists in Qing China: Science, Empire, and Cultural Encounter*. Cambridge, MA: Harvard University Press, 2004.

———. "The Global Turn in the History of Science." *East Asian Science, Technology and Society* 6 (2012): 249–58.

Furth, Charlotte. "The Physician as Philosopher of the Way: Zhu Zhenheng (1282–1358)." *Harvard Journal of Asiatic Studies* 66, no. 2 (2006): 423–59.

Goldschmidt, Asaf. *The Evolution of Chinese Medicine: Song Dynasty, 960–1200*. New York: Routledge, 2008.

Han'guk chŏngsin munhwa yŏn'guwŏn 한국정신문화연구원. *Han'guk minjok munhwa tae paekkwa sajŏn* 한국민족문화대백과사전 (The great encyclopedia of Korean nation and culture). 27 volumes. Seoul: Ungjin, 1991.

Han'guk insamsa p'yŏnch'an wiwŏnhoe 한국인삼사편찬위원회. *Han'guk insamsa* 한국인삼사 (A history of Korean ginseng). Seoul: Tongil munhwasa, 2002.

Han'guk kwanggo yŏn'guwŏn 한국광고연구원. *Han'guk kwanggo paengnyŏn* 한국광고백년 (Korean advertising: 100 years). Seoul: Han'guk kwanggo tanch'e yŏnhaphoe, 1996.

Hanson, Marta. *Speaking of Epidemics in Chinese Medicine: Disease and the Geographic Imagination in Late Imperial China*. London: Routledge, 2011.

Hanyu da zidian 漢語大字典 (The great dictionary of Chinese). Wuhan, China: Hubeicishu chubanshe, 1990.

Haynes, Douglas E. "Creating the Consumer? Advertising, Capitalism, and the Middle Class in Urban Western India, 1914–40." In *Towards a History of Consumption in South Asia*, edited by Douglas E. Haynes and Abigail McGowan, 185–223. Oxford: Oxford University Press, 2010.

Henry, Todd A. "Sanitizing Empire: Japanese Articulations of Korean Otherness and the Construction of Early Colonial Seoul, 1905–1919." *Journal of Asian Studies* 64 (2005): 639–75.

Hinrichs, T. J., and Linda L. Barnes, eds. *Chinese Medicine and Healing: An Illustrated History*. Cambridge, MA: Harvard University Press, 2013.

Holmes, Thomas H., and Richard H. Rahe. "The Social Readjustment Rating Scale." *Journal of Psychosomatic Research* 11, no. 2 (1967): 213–18.

Hong Hyŏn-o 홍현오. *Han'guk yagŏpsa* 한국약업사 (A Korean history of pharmacy). Seoul: Handok yagŏp chusikhoesa, 1972.

Hong Yun-p'yo 홍윤표. "Mulmyŏnggo e taehan koch'al" 물명고에대한고찰 (A study on *Mulmyeongo*). *Chindan hakpo* 118 (2013): 167–211.

Hughes, Theodore, Jae-young Kim, Jin-kyung Lee, and Sang-kyung Lee, eds. *Rat Fire: Korean Stories from the Japanese Empire*. Ithaca, NY: Cornell University Press, 2013

Hwang, Kyung Moon. *Beyond Birth: Social Status in the Emergence of Modern Korea*. Cambridge, MA: Harvard University Press, 2004.

Hwang Ŭi-wan 황의완. *Hwabyŏng kŭkpok p'ŭrojekt'ŭ* 화병극복프로젝트 (A project to overcome *hwabyŏng*). Seoul: Chosŏn maegŏjin, 2011.

Imamura Tomo 今村鞆. *Ninjinshi* 人蔘史 (A history of ginseng). 7 vols. Keijō: Chōsen sōtokofu senbaikyoku, 1934–1940.

Johnston, William. *The Modern Epidemic*. Cambridge, MA: Harvard University Press, 1995.

Kang Chu-bong 강주봉, Kwŏn Sun-jong 권순종, and Ch'oe Chun-bae 최준배. *Imsangŭi rŭl wihan sanghannon kangjwa* 임상의를위한상한론강좌 (A lecture for practitioners: On *Treatise on Cold-Damage Disorders*). Seoul: Minjok ŭihak sinmunsa, 2010.

Kang Kyŏng-ae 강경애. "Sogŭm" 소금 (Salt). 1934. Translated by Jin-kyung Lee. In *Rat Fire: Korean Stories from the Japanese Empire*, edited by Theodore Hughes, Jae-young Kim, Jin-kyung Lee, and Sang-kyung Lee, 215–62. Ithaca, NY: Cornell University Press, 2013.

Kang Yong-wŏn 강용원. *Annyŏng, uulchŭng* 안녕우울증 (Good-bye, depression). Seoul: Miraerŭl soyuhan saramdŭl, 2011.

Karchmer, Eric. "Chinese Medicine in Action: On the Postcoloniality of Medical Practice in China." *Medical Anthropology* 29, no. 3 (2010): 226–52.

Kawakami Takeshi 川上武. *Gendai Nihon iryōshi* 現代日本医療史 (A history of modern Japanese medicine). Tokyo: Keisō shobō, 1965.

Ki Ch'ang-dŏk 기창덕. "Ŭihak kyoyuk ŭi hyŏndaehwa kwajŏng" 의학교육의현대화과정 (Modernization process of the medical education in Korea). *Ŭisahak* 3, no. 1 (1994): 72–129.

Kim Chong-u 김종우. *Hwabyŏng ŭro put'ŏ ŭi haebang* 화병으로부터의해방 (Liberation from *hwabyŏng*). Seoul: Yŏsŏng sinmunsa, 2007.

Kim Chong-u 김종우 and Hwang Ŭi-wan 황의완. "Hanŭihak esŏ pon hwabyŏng ŭi haesŏk" 한의학에서본화병의해석 (*Hwabyung* in the view of Oriental medicine). *Tongŭi sin'gyŏng chŏngsin kwahak hoeji* 5, no. 1 (1994): 9–15.

———. "Hwabyŏng hwanja ŭi hanŭihakchŏk ch'iryo e taehan imsangjŏk yŏn'gu" 화병환자의한의학적치료에대한임상적연구 (A clinical study on treatments of *hwabyung* with Oriental medicine). *Taehan hanŭihak hoeji* 19, no. 2 (1998): 5–16.

Kim Chong-u 김종우, Chŏng Sŏn-yong 정선용, Cho Sŏng-hun 조성훈, Hwang Ŭi-wan 황의완, and Kim Po-gyŏng 김보경. "Hwabyŏng imsang chillyo chich'im kaebal yŏn'gu (1)-mokchŏk kwa kaebal chŏlyak mit chŏlch'a" 화병임상진료지침개발연구(1)-목적과개발전략및절차 (Development of clinical practice guideline for *hwabyung* (1): Purpose, development strategy and procedure). *Tongŭi sin'gyŏng chŏngsin kwahak hoeji* 20, no. 2 (2009): 143–52.

Kim Chong-u 김종우, Kim Sang-ho 김상호, Chŏng Sŏn-yong 정선용, Pak So-jŏng 박소정, Pyŏn Sun-im 변순임, Kim Chi-yŏng 김지영, and Hwang Ŭi-wan 황의완. "Hwabyŏng kwa hwabyŏng chuyo uulchŭng chungbok chindan ŭi OMS-prime ŭl t'onghan pyŏngjŭng yuhyŏng pigyo yŏn'gu" 화병과화병주요우울증증복진단의 OMS-prime 을통한병증유형비교연구 (A comparative study on pattern identification by OMS-prime of *hwa-byung* group and *hwa-byung* with major depression double diagnosis group). *Tongŭi sin'gyŏng chŏngsin kwahak hoeji* 18, no. 3 (2007): 1–14.

Kim Chong-u 김종우, Hyŏn Kyŏng-ch'ŏl 현경철, and Hwang Ŭi-wan 황의완. "Hwabyŏng ŭi kiwŏn e kwanhan koch'al-Chosŏn wangjo sillok ŭl chungsimŭro" 화병의기원에관한고찰-조선왕조실록을중심으로 (A study on the origin of *hwa-byung* based on the *Veritable Records of Chosun Korea*). *Tongŭi sin'gyŏng chŏngsin kwahak hoeji* 10, no. 1 (1999): 205–17.

Kim Chong-u 김종우, Chŏng Sŏn-yong 정선용, Sŏ Hyŏn-uk 서현욱, Chŏng In-ch'ŏl 정인철, Yi Sŭng-gi 이승기, Kim Po-gyŏng 김보경, Kim Kŭn-u 김근우, Yi Chae-hyŏk 이재혁, Kim Nak-hyŏng 김낙형, Kim T'ae-hŏn 김태헌, and Kang Hyŏng-wŏn 강형원. "Hwabyŏng yŏkhak yŏn'gu charyo rŭl kibanŭro han hwabyŏng hwanja ŭi t'ŭksŏng" 화병역학연구자료를기반으로한화병환자의특성 (The characteristics of *hwa-byung* patients based on *hwa-byung* epidemiologic data). *Tongŭi sin'gyŏng chŏngsin kwahak hoeji* 21, no. 2 (2010): 157–69.

Kim, Hee-soo. "Roots of Han and Its Healing: A Study of Han from the Perspective of Christian Ethics." Ph.D. diss., Graduate Theological Union, 1994.

Kim Ho 김호. "Chŏngjodae ŭiryo chŏngch'aek" 정조대의료정책 (Medical policy during the reign of King Chŏngjo). *Han'guk hakpo* 22, no. 1 (1996): 231–57.

——. *Hŏ Chun ŭi "Tongŭi pogam" yŏn'gu* 허준의동의보감연구 (A study on Hŏ Chun's *Tongŭi pogam*). Seoul: Ilchisa, 2000.

Kim, Hoi-Eun. "Anatomically Speaking: The Kubo Incident and the Paradox of Race in Colonial Korea." In *Race and Racism in Modern East Asia: Western and Eastern Constructions*, edited by Rotem Kowner and Walter Demel, 411–30. Leiden: Brill, 2013.

——. *Doctors of Empire: Medical and Cultural Encounters between Imperial Germany and Meiji Japan*. Toronto: University of Toronto Press, 2014.

Kim Hyŏng-sŏk 김형석. "Hanmal Han'gugin e ŭihan sŏyang ŭihak suyong" 한말한 국인에의한서양의학수용 (The Korean accommodation of Western medicine in late Chosŏn). *Kuksagwan nonch'ong* 5 (December 1989): 171–210.

Kim Kŭn-bae 김근배. "Ilche kangjŏmgi Chosŏnindŭl ŭi ŭisa toegi" 일제강점기조선 인들의의사되기 (Becoming medical doctors in colonial Korea). *Ŭisahak* 23, no. 3 (2014): 429–68.

Kim Kyo-sik 김교식. *Yu Ir-han, Yuhan Yanghaeng gŭrup* 유일한유한양행그룹 (Yu Ir-han and the Yuhan Company). Seoul: Kyesŏng ch'ulp'ansa, 1984.

Kim Nam-il 김남일. "Han'guk sanghannon yŏn'gu yaksa" 한국상한론연구약사 (A brief history of study on *Shanghanlun* in Korea). Paper presented at "Chang Chunggyŏng haksul taehoe" 장중경학술대회, Seoul, October 22–27, 2009.

——. *Hanŭihak e mich'in Chosŏn ŭi chisigindŭl: yuŭi yŏlchŏn* 한의학에미친조선의 지식인들: 유의열전 (The Chosŏn intellectuals obsessed by traditional medicine). Seoul: Tŭllyŏk, 2011.

Kim Ok-chu 김옥주. "Kyŏngsŏng chedae ŭihakpu ŭi ch'ejil illyuhak yŏn'gu" 경성제 대의학부의체질인류학연구 (Physical anthropology studies at Keijō Imperial University Medical School). *Ŭisahak* 17 (2008): 191–203.

Kim Sin-gŭn 김신근. *Han'guk ŭiyaksa* 한국의약사 (Affairs of medicine and pharmacology in Korea). Seoul: Seoul National University, 2001.

Kim, Sonja. "Contesting Bodies: Managing Population, Birthing, and Medicine in Korea, 1876–1945." Ph.D. diss., University of California, Los Angeles, 2008.

Kim Sŭng-t'ae 김승태. "Ilbon ŭl t'onghan sŏyang ŭihak ŭi suyong kwa kŭ sŏnggyŏk" 일본을통한서양의학의수용과그성격 (The attributes of Korean accommodation of Western medicine via Japan). *Kuksagwan nonch'ong* 6 (December 1989): 223–54.

Kim, Sun Joo. *Marginality and Subversion in Korea: The Hong Kyŏngnae Rebellion of 1812*. Seattle: University of Washington Press, 2007.

Kim Tae-wŏn 김대원. "18 segi min'gan ŭiryo ŭi sŏngjang" 18세기민간의료의성장 (The growth of private medicine in the eighteenth century). Master's thesis, Seoul National University, 1998.

Kim Tu-jong 김두종. *Han'guk ŭihaksa* 한국의학사 (A history of Korean medicine). Seoul: Tamgudang, 1966.

Kim Yŏl-gyu 김열규. *Han'gugin ŭi hwa* 한국인의화 (The *hwa* of Koreans). Seoul: Hyumŏnisŭt'ŭ 2004.

Kim, Yung Sik. "'The Problem of China' in the Study of the History of Korean Science: Korean Science, Chinese Science, and East Asian Science." *Gujin lunheng* 18 (2008): 185–98.

———. "Problems and Possibilities in the Study of the History of Korean Science." *Osiris* 13 (1998): 48–79.

Kleinman, Arthur. *The Illness Narratives: Suffering, Healing, and the Human Condition*. New York: Basic Books, 1988.

Ko Ŭn 고은. "Han ŭi kŭkpok ŭl wihayŏ" 한의극복을위하여 (To overcome *han*). In *Han ŭi iyagi* 한의이야기 (The story of *han*), edited by Sŏ Kwang-sŏn 서광선, 23–59. Seoul: Pori, 1988.

Kuriyama, Shigehisa. *The Expressiveness of the Body and the Divergence of Greek and Chinese Medicine*. New York: Zone Books, 1999.

Kwŏn Pyŏng-t'ak 권병탁. *Yangnyŏngsi yŏn'gu* 약령시연구 (A study of herbal markets). Seoul: Han'guk yŏn'guwon, 1986.

Latour, Bruno. *We Have Never Been Modern*. Translated by Catherine Porter. Cambridge, MA: Harvard University Press, 1993.

Lee, Ki-baik. *A New History of Korea*. Translated by Edward W. Wagner with Edward J. Shultz. Seoul: Ilchogak, 1984.

Lei, Sean Hsiang-lin. "From Changshan to a New Anti-malarial Drug: Re-networking Chinese Drugs and Excluding Traditional Doctors." *Social Studies of Science* 29, no. 3 (1999): 323–58.

Lewis, Martin W., and Kären E. Wigen. *The Myth of Continents: A Critique of Metageography*. Berkeley: University of California Press, 1997.

Li Jianmin 李建民. "'Bencao gangmu. huobu' kaoshi" 本草綱目.火部"考釋 (Fire as medicine: The "fire" section of the *Bencao gangmu*). *Zhongyang yanjiuyuan lishiyuyan yanjiusuo jikan* 73, no. 3 (2002): 395–441.

Liang Yongxuan 梁永宣. "Chaoxian 'Yilin cuoyao' suozai zhongchao yixue jiaoliu shiliao yanjiu" 朝鮮"醫林撮要"所載中朝醫學交流史料研究 (*Essentials of Medical Works* and its historical materials of medical exchanges between China and Korea). *Zhonghua yishi zazhi* 31, no. 1 (January 2001): 17–20.

———. "Wang Yinglin yu 'Da chaoxian yi wen'" 王應遴與"答朝鮮醫問 (Wang Yinglin and his *Answer to the Questions of a Korean Physician*). *Zhonghua yishi zazhi* 30, no. 2 (April 2000): 69–72.

Lin, Keh-Ming. "Hwa-byung: A Korean Culture-Bound Syndrome?" *American Journal of Psychiatry* 140 (1983): 105–7.

Lin, Keh-Ming, John K.C. Lau, Joe Yamamoto, Yan-Ping Zheng, Hun-soo Kim, Kyu-hyoung Cho, and Gayle Nakasaki. "*Hwa-byung*: A Community Study of Korean-Americans." *Journal of Nervous Mental Disease* 180, no. 6 (1992): 386–91.

Liu Yanan 劉亞男. "Huang Yuanyu shengping ji qi heshu gongxien chutan" 黃元御生平及其學術貢獻初探 (A preliminary study of Huang Yuanyu's life and his academic contribution). Master's thesis, Beijing zhongyiyao daxue, 2012.

Lloyd, Camille. "Life Events and Depressive Disorder Reviewed: Events as Predisposing Factors." *Archives of General Psychiatry* 37, no. 5 (1980): 529–35.

Lloyd, G. E. R., ed. *Hippocratic Writings*. Translated by J. Chadwick and W. N. Mann. London: Penguin Books, 1983.

Lo, Ming-Cheng M. *Doctors within Borders: Profession, Ethnicity, and Modernity in Colonial Taiwan*. Berkeley: University of California Press, 2002.

Ma, Eun Jeong. "The Medicine Cabinet: Korean Medicine under Dispute." *East Asian Science Technology and Society* 4, no. 3 (2010): 367–82.

Makeham, John. *Name and Actuality in Early Chinese Thought*. Albany: State University of New York Press, 1994.

Mayers, Linda. "A Commentary on Myra's Life: The Creation of a Story." *Psychoanalytic Inquiry* 20 (2010): 63–70.

McKeown, Thomas. *The Modern Rise of Population*. New York: Academic Press, 1976.

Miki Sakae 三木榮. *Chōsen igakushi oyobi shippeishi* 朝鮮醫學史及疾病史 (History of Korean medicine and of disease in Korea). Osaka: Shibun chuppansha, 1962.

Min Sŏng-gil [Min, Sung Kil] 민성길. "Differences in Temperament and Character Dimensions of Personality between Patients with Hwa-byung, an Anger Syndrome, and Patients with Major Depressive Disorder." *Journal of Affective Disorders* 138, no. 2 (2012): 110–16.

———. *Hwabyŏng yŏn'gu* 화병연구 (A study of *hwa-byung*). Seoul: Emel k'ŏmyunik'eisyŏn, 2009.

———. "Hwa-Byung: An Anger Syndrome and Proposing New Anger Disorder." In *Psychology of Anger: Symptoms, Causes and Coping*, edited by James P. Welty, 1–49. Hauppage, NY: Nova Science, 2011.

———. "Hwatpyŏng kwa han" 홧병과한 (*Hwabyŏng* and the Psychology of Han). *Taehan ŭihak hyŏphoeji* 34, no. 11 (1991): 1189–98.

Min Sŏng-gil 민성길, So Ŭn-hŭi 소은희, and Pyŏn Yong-uk 변용욱. "Chŏngsin'gwa ŭisa wa hanŭisa tŭl ŭi hwatpyŏng e taehan kaenyŏm" 정신과의사와한의사들의홧병에대한개념 (The concept of *hwa-byung* of Korean psychiatrists and herb physicians). *Sin'gyŏng chŏngsin ŭihak* 28, no. 1 (1989): 146–54.

Min Sŏng-gil 민성길, Sŏ Sin-yŏng 서신영, Cho Yun-gyŏng 조윤경, Hŏ Chi-ŭn 허지은, and Song Ki-jun 송기준. "Hwabyŏng ch'ŏkto wa yŏn'guyong hwabyŏng chindan kijun kaebal" 화병척도와연구용화병진단기준개발 (Development of *hwa-byung* scale and research criteria of *hwa-byung*). *Sin'gyŏng chŏngsin ŭihak* 48, no. 2 (2009): 77–85.

Min Sŏng-gil 민성길, Yi Man-hong 이만홍, Sin Chŏng-ho 신정호, Pak Muk-hŭi 박묵희, Kim Man-gwŏn 김만권, and Yi Ho-yŏng 이호영. "Hwatpyŏng e taehan chindanjŏk yŏn'gu" 홧병에대한진단적연구 (A diagnostic study on *hwabyung*). *Taehan ŭihak hyŏphoeji* 29, no. 6 (1986): 653–61.

Min Sŏng-gil 민성길 and Kim Chin-hak 김진학. "Pogilto esŏ ŭi hwabyŏng e taehan yŏn'gu" 보길도에서의화병에대한연구 (A study on *hwabyung* in Bokil Island). *Sin'gyŏng chŏngsin ŭihak* 25, no. 3 (1986): 459–66.

Morohashi Tetsuji 轍次諸橋. *Dai kan-wa jiten* 大漢和辞典 (Great Chinese-Japanese dictionary). 13 vols. Tokyo: Taishūkan Shoten Takasuke Monohashi, 1990.

Morris-Suzuki, Tessa. "Debating Racial Science in Wartime Japan." *Osiris* 13 (1998): 354–75.

Mukharji, Projit B. *Nationalizing the Body: The Medical Market, Print and Daktari Medicine*. London: Anthem, 2009.

Mun Chung-yang 문중양. "Sejongdae kwahak kisul ŭi chajusŏng tasi pogi" 세종대과학기술의'자주성'다시보기 (Science and technology in King Sejong's era; questioning their independence from the Chinese ones). *Yŏksa hakpo* 189 (2006): 39–72.

Nam P'ung-hyŏn 남풍현. "Hyangyak chipsŏngbang ŭi hyangmyŏng e taehayŏ" 향약집성방의향명에대하여 (On the local names of the *Standard Prescriptions of Local Botanicals*). *Chindan hakpo* 87 (1999): 171–94.

Nappi, Carla. *The Monkey and the Inkpot*. Cambridge, MA: Harvard University Press, 2009.

Needham, Joseph. *Science and Civilization in China*. Vol. 7. Cambridge: Cambridge University Press, 2004.

O Chae-gŭn 오재근. "Chosŏn ŭisŏ 'Hyangyak chipsŏngbang' chung e sillin sanghan nonŭi yŏn'gu" 조선의서향약집성방중에실린상한논의연구 (A study on cold damage in the *Conpendium of Prescription from the Countryside*). *Han'guk ŭisahak hoeji* 25, no. 2 (2012): 121–36.

Oh, Se-mi. "Consuming the Modern: The Everyday in Colonial Seoul, 1915–1937," Ph.D. diss., Columbia University, 2008.

Paek In-je paksa chŏn'gi kanhaeng wiwŏnhoe 백인제박사전기간행위원회. *Sŏngakcha Paek In-je* 선각자백인제 (Paek In-je, the pioneer). Seoul: Ch'angjak kwa pip'yŏngsa, 1999.

Pak Hyŏng-u 박형우. "Taeŭi Kim P'il-sun" 대의김필순 (Kim Pil-soon, a great doctor). *Ŭisahak* 7, no. 2 (1998): 239–53.

Pak Myŏng-gyu 박명규. *Singmin kwŏllyŏk kwa t'onggye—Chosŏn ch'ongdokpu ŭi t'onggye ch'eje wa sensŏsŭ* 식민지권력과통계 (Colonial authority and statistics). Seoul: Seoul National University Press, 2003.

Pak Sŏng-sik 박성식. "Tongmu Yi Che-ma ŭi kagye wa saengae e taehan yŏn'gu" 동무이제마의가계와생애에대한연구 (A study on Yi Che-ma's family lineage and life). *Sasang ŭihak hoeji* 8, no. 1 (1996): 17–32.

Pak Sŏng-sik 박성식 and Song Il-byŏng 송일병. "Sasang ŭihak ŭi haksulchŏk yŏnwŏn kwa Yi Che-ma ŭihak sasang e kwanhan yŏn'gu" 사상의학의학술적연원과이제마의학사상에관한연구 (A study on the intellectual origins of Four Constitutions medicine and Yi Che-ma's medical thoughts). *Sasang ŭihak hoeji* 5, no. 1 (1993): 7–39.

Pak Sun-yŏng 박순영. "Ilche singmin chuŭi wa Chosŏnin ŭi mom e taehan 'illuhakchŏk sisŏn'" 일제식민주의와조선인의몸에대한"인류학적"시선 (The "anthropological" gaze at the Korean bodies under Japanese colonialism). *Pigyo munhwa yŏn'gu* 12, no. 2 (2006): 76–78.

———. "Ilche singmin t'ongch'i ha ŭi Chosŏn ch'ejil illuhak i namgin hangmunjŏk kwaje wa sŏgu ch'ejil illuhaksa ro put'ŏ ŭi kyohun" 일제식민통치하의조선인체질인류학이남긴학문적과제와서구체질인류학사로부터의교훈 (Academic tasks left by physical anthropology under Japanese colonialism and lessons from experiences of Western physical anthropology). *Pigyo munhwa yŏn'gu* 10, no. 1 (2004): 191–220.

Pak Yun-jae 박윤재. "Ch'ŏngsim pomyŏngdan nonjaeng e panyŏngdoen t'onggambu ŭi ŭiyakp'um chŏngch'aek" 청심보명단논쟁에반영된통감부의의약품정책 (The residency-general's policy of medicinal goods viewed from the controversy over Ch'ŏngsim pomyŏngdan). *Yŏksa pip'yŏng* 67 (2004): 191–206.

———. *Han'guk kŭndae ŭihak ŭi kiwŏn* 한국근대의학의기원 (The origins of modern medicine in Korea). Seoul: Hyean, 2005.

Pang, Keum Young C. "Hwabyung: The Construction of a Korean Popular Illness among Korean Elderly Immigrant Women in the United States." *Culture, Medicine, and Psychiatry* 14 (1990): 495–512.

Park, Jin-Kyung. "Corporeal Colonialism: Medicine, Reproduction, and Race in Colonial Korea." Ph.D. diss., University of Illinois at Urbana-Champaign, 2008.

Pollock, Sheldon. "Cosmopolitan and Vernacular in History." *Public Culture* 12, no. 3 (2000): 591–625.

Porter, Roy. "'Expressing Yourself Ill': The Language of Sickness in Georgian England." In *Language, Self, and Society: A Social History of Language*, edited by Peter Burke and Roy Porter, 276–299. Cambridge: Polity, 1991.

Raj, Kapil. *Relocating Modern Science: Circulation and the Construction of Knowledge in South Asia and Europe, 1650–1900.* New York: Palgrave Macmillan, 2007.

Reichert, Julie, Brian Solan, Craig Timm, and Summers Kalishman. "Narrative Medicine and Emerging Clinical Practice." *Literature and Medicine* 27, no. 2 (2008): 248–71.

Rogaski, Ruth. "Vampires in Plagueland: The Multiple Meanings of *Weisheng* in Manchuria." In *Health and Hygiene in Chinese East Asia*, edtited by Angela Ki Che Leung and Charlotte Furth, 132–59. Durham, NC: Duke University Press, 2010.

Safier, Neil. "Global Knowledge on the Move: Itineraries, Amerindian Narratives, and Deep Histories of Science." *Isis* 101, no. 1 (2010): 133–45.

Samman, Khaldoun. "The Limits of the Classical Comparative Method: The Search for an Alternative Method for Studying 'Other' Societies, Cultures, and Civilizations." *Review: Journal of the Fernand Braudel Center for the Study of Economics, Historical Systems, and Civilizations* 24, no. 4 (2001): 533–73.

Satō Gōzō 佐藤剛藏. *Chosen yiikushi* 朝鮮医育史 (A history of medical education in Chosŏn Korea). Translated into Korean by Yi Ch'ung-ho. *Chosŏn ŭiyuksa*. Seoul: Hyŏngsŏl ch'ulp'ansa, 1992.

Scheid, Volker. "Constraint 鬱 as a Window on Approaches to Emotion-Related Disorders in East Asian Medicine." *Culture, Medicine, and Psychiatry* 37 (2013): 2–7.

Simonis, Fabien. "Mad Acts, Mad Speech, and Mad People in Late Imperial Chinese Law and Medicine." Ph.D. diss., Princeton University, 2010.

Sin In-sŏp 신인섭. *Han'guk kwanggo paltalsa* 한국광고발달사 (A history of advertising and its development). Seoul: Ilchogak, 1980.

———. "Han'guk ŭi kukche kwanggosa; Sŏul p'ŭresŭ e sillin yuhan yanghaeng kwanggo" 한국의국제광고사 (A history of international advertisements in Korea). *Kwanggo Chŏngbo* 263 (February 2003): 68–73.

Sin In-sŏp 신인섭 and Sŏ Pŏm-sŏk 서범석. *Han'guk kwanggosa* 한국광고사 (A history of advertising and its development). Seoul: Nanam, 2011.

Sin Tong-wŏn [Shin, Dongwon] 신동원. *Chosŏn saram Hŏ Chun* 조선사람허준 (Hŏ Chun, the Chosŏn Korean). Seoul: Han'gyŏre sinmunsa, 2001.

——. *Chosŏn saram ŭi saengno pyŏngsa* 조선사람의생로병사 (The birth, old age, sickness, and death of Chosŏn Koreans). Seoul: Han'gyŏre ch'ulp'ansa, 1999.

——. "How Four Different Political Systems Have Shaped the Modernization of Traditional Korean Medicine between 1900 and 1960." *Historia Scientiarum* 17, no. 3 (2008): 225–41.

——. *Hoyŏlcha, Chosŏn ŭl sŭpkyŏk hada* 호열자조선을습격하다 (Hoyŏlcha attacked Chosŏn Korea). Seoul: Yŏksa pip'yŏngsa, 2004.

——. "Hyangyak ŭisul i in'gu rŭl chŭngga sik'yŏssŭlkka" 향약의술이인구를증가시켰을까 (Did the medicine of local botanicals increase population?). *Yŏksa pip'yŏng* 61 (2002): 251–64.

——. "Korean Anatomical Charts in the Context of the East Asian Medicine." *Asian Medicine* 5 (2009): 186–207.

——. "Nationalistic Acceptance of Sasang Medicine." *Review of Korean Studies* 9, no. 2 (2006): 143–63.

——. *"Tongŭi pogam" kwa tongasia ŭihaksa* 동의보감과동아시아의학사 (*Donguibogam* and the history of medicine in East Asia). Seoul: Tŭllyŏk, 2015.

——. "Yuŭi ŭi kil: Chŏng Yag-yong ŭi ŭihak kwa ŭisul" 유의의길:정약용의의학과의술 (Cheong Yagyong's medical thought and practice: The whole picture and its development). *Tasanhak* 10 (2007): 171–224.

Sivasundaram, Sujit. "Global Histories of Science." *Isis* 101, no. 1 (2010): 95–97.

——. "Sciences and the Global: On Methods, Questions, and Theory." *Isis* 101, no. 1 (2010): 146–58.

Sŏ Hong-gwan 서홍관 and Sin Chwa-sŏp 신좌섭. "Ilbon injongnon kwa Chosŏnin" 일본인종론과조선인 (Japanese ethnology and Chosŏn people during Japanese annexation period). *Ŭisahak* 8, no. 1 (1999): 59–68.

Son, Chang-hee. *Haan of Minjung Theology and Han of Han Philosophy*. Lanham, MD: University Press of America, 1984.

Song Kyŏng-sŏp 송경섭. "Hanŭihak ŭi palchŏn kwajŏng kwa sahoe kiyŏdo e kwanhan yŏn'gu" 한의학의발전과정과사회기여도에관한연구 (A study on the developmental process and social contribution of Oriental medicine in Korea). Ph.D. diss., Kyŏngsan University, 1994.

Tashiro Kazui 田代和生. *Wakan: sakoku jidai no Nihonjin-machi* 倭館: 鎖国時代の日本人町 (Wakan: The city of Japanese in the period of national isolation). Translated into Korean by Chŏng Sŏng-il. *Waegwan: Chosŏn ŭn wae ilbon saramdŭl ŭl kaduŏssŭlkka?* Seoul: Nonhyŏn, 2005.

Terazawa, Yuki. "Gender, Knowledge, and Power: Reproductive Medicine in Japan, 1790–1930." Ph.D. diss., University of California, Los Angeles, 2001.

Tilley, Helen. "Global Histories, Vernacular Science, and African Genealogies; or, Is the History of Science Ready for the World?" *Isis* 101, no. 1 (2010): 110–19.

T'onggyech'ŏng 통계청. "Samang wŏnin t'onggye" 사망원인통계 (Causes of death statistics), 2003–2014.

Trambaiolo, Daniel. "Diplomatic Journeys and Medical Brush Talks: Eighteenth-Century Dialogues between Korean and Japanese Medicine." In *Orbits, Routes and Vessels: Motion and Knowledge in the Changing Early Modern World*, edited by Ofer Gal and Yi Zheng, 93–113. Dordrecht, Netherlands: Springer, 2014.

Unschuld, Paul U. *Medicine in China: A History of Ideas*. Berkeley: University of California Press, 1985.

———. *Medicine in China: A History of Pharmaceutics*. Berkeley: University of California Press, 1986.

Wales, Nym [Snow, Helen F.], and Kim San. *Song of Ariran: A Korean Communist in the Chinese Revolution*. San Francisco: Ramparts, 1941.

Walker, Timothy D. "The Medicines Trade in the Portuguese Atlantic World: Acquisition and Dissemination of Healing Knowledge from Brazil (c. 1580–1800)." *Social History of Medicine* 26, no. 3 (2013): 403–31.

Watters, Ethan. *Crazy Like Us: The Globalization of the American Psyche*. New York: Free Press, 2010.

Willoughby, Heather A. "The Sound of *Han*: Pansori, and a South Korean Discourse of Sorrow and Lament." Ph.D. diss., Columbia University, 2002.

Wu, Yi-Li. *Reproducing Women: Medicine, Metaphor, and Childbirth in Late Imperial China*. Berkeley: University of California Press, 2010.

Yamada Keiji 山田慶兒. *Chugoku igaku wa ikani tsukuraretaka* 中国医学はいかにつくられたか (How was Chinese medicine constructed?). Translated into Korean by Chŏn Sang-un and Yi Sŏng-gyu. *Chungguk ŭihak ŭn ŏttŏk'e sijak toeŏnnŭn'ga?* Seoul: Saiŏnsŭbuksŭ, 1999.

Yang Chŏng-p'il 양정필. "Hanmal ilche ch'o kŭndaejŏk yagŏp hwan'gyŏng ŭi taedu wa hanyagŏpcha ŭi taeŭng" 한말일제초근대적약업환경의대두와한약업자의대응 (Modern medicine environment and adaptation of Korean trader for medicinal herbs from the late nineteenth century to the early twentieth century). *Ŭisahak* 15, no. 2 (2006): 189–209.

Ye Chong-sŏk 예종석. *Hwalmyŏngsu 100 nyŏn sŏngjang ŭi pimil* 활명수 100년성장의비밀 (Hwalmyŏngsu, the secret of 100 years of development). Seoul: Lidŏsŭbuk, 2009.

Yi Hŭi-yŏng 이희영, Kim Chong-u 김종우, Pak Chong-hun 박종흔, and Hwang Ŭi-wan 황의완. "Hwabyŏng ŭi chindan mit pyŏngjŭng yuhyŏng e kwanhan yŏn'gu" 화병의진단및병증유형에관한연구 (A study for diagnosis and pattern identification of *hwa-byung*). *Tongŭi sin'gyŏng chŏngsin kwahak hoeji* 16, no. 1 (2005): 1–17.

Yi Ki-bok 이기복. "18 segi ŭigwan Yi Su-gi ŭi chagi insik: kisulchik chungin ŭi chŏnmunga ŭisik ŭl chungsimŭro" 18세기의관이수기의자기인식:기술직중인의전문가의식을중심으로 (Yi Suki's *Yŏksi manp'il* and the professional identity of a Chungin medical official in eighteenth-century Chosŏn Korea). *Ŭisahak* 22, no. 2 (2013): 483–528.

———. "Tongmu Yi Che-ma ŭi ŭihak sasang kwa silch'ŏn" 동무이제마의의학사상과실천 (Restructuring East Asian medical tradition: The medical theory and practice of Yi Che-ma in late nineteenth-century Korea). Ph.D. diss., Seoul National University, 2014.

———. "Ŭian ŭro salp'yŏ ponŭn Chosŏn hugi ŭi ŭihak" 의안으로살펴보는조선후기의의학 (Eighteenth- and nineteenth-century Korean medicine viewed through medical case writings). *Han'guk kwahaksahak hoeji* 34, no. 3 (2012): 429–58.

Yi Kyŏng-nim 이경림. "Korean Women." In *Echoing Song: Contemporary Korean Women Poets*, edited by Peter H. Lee, 96. Buffalo, NY: White Pine Buffalo, 2005.

———. *A New Season Approaching: Devour It*. Translated by Wolhee Choe. New York: Hawks, 2011.

Yi Kyŏng-nok 이경록. "Chosŏn chŏn'gi 'Ŭibang yuch'wi' ŭi sŏngch'wi wa han'gye" 조선전기"의방유취"의성취와한계 (The achievement and limitations of the *Euibangyoochui* during the early days of the Joseon). *Han'guk kwahaksahak hoeji* 34, no. 3 (2012): 462–92.

———. "'Hyangyak chipsŏngbang' ŭi p'yŏnch'an kwa Chungguk ŭiryo ŭi Chosŏnhwa" 향약집성방의편찬과중국의료의조선화 (The publication of *Hyangyak jipseongbang* and the Chosŏnization of the Chinese medicine). *Ŭisahak* 20, no. 2 (2011): 225–62.

———. *Koryŏ sidae ŭiryo ŭi hyŏngsŏng kwa palchŏn* 고려시대의료의형성과발전 (The formation and development of medicine during the Koryŏ dynasty). Seoul: Hyean, 2010.

Yi Kyu-sik 이규식. "Ilche ŭi nongch'on ch'imt'al kwa nongch'on wisaeng yŏn'guso" 일제의농촌침탈과농촌위생연구소 (A study of the pillages of the Korean rural villages under the rule of Japanese imperialism and the Research Institute for Rural Health). *Ŭisahak* 10, no. 2 (2001): 124–34.

———. "Yu Il-chun ŭi saengae wa hangmun" 유일준의생애와학문 (A study about Il Chun Yu). *Ŭisahak* 12, no. 1 (2003): 1–12.

Yi Sang-wŏn 이상원. "'Tongŭi pogam' ŭi sanghan e taehan insik" 동의보감의상한에대한인식 (A study of *Dongeuibogam*'s sanghan). Master's thesis, Kyung Hee University, 2004.

Yi Si-hyŏng 이시형. "Hwabyŏng ŭi kaenyŏm e taehan yŏn'gu" 화병의개념에대한연구 (A study on the concept of *hwabyŏng*). *Koryŏ pyŏngwŏn chapchi* 1 (1977): 63–69.

Yi Sŏn-a 이선아. "19segi koch'ang chibang ŭiwŏn Ŭn Su-ryong i namgin kyŏnghŏm ŭian" 19세기고창지방의원은수룡이남긴경험의안 (A study of the case records of an herbalist En Su-ryong in the late Chosun dynasty). *Han'guk ŭisahak hoeji* 18, no. 2 (2005): 63–91.

Yi Sŏn-a 이선아 and Yi Si-hyŏng 이시형. "Hwang To-yŏn ŭi 'Pangyak happyŏn' e kwanhan yŏn'gu" 황도연의"방약합편"에관한연구 (A study of *PangYak-Happyon*). *Han'guk chŏnt'ong ŭihakji* 11, no. 1 (2001): 101–9.

Yi T'ae-jin 이태진. "Ohae wa yihae pujok ŭi chipchung punsŏk" 오해와이해부족의집중분석 (A critical analysis of misunderstanding and the lack of understanding). *Yŏksa pip'yŏng* 62 (2003): 225–34.

———. *Ŭisul kwa in'gu kŭrigo nongŏp kisul* 의술과인구그리고농업기술 (Medicine, population, and agricultural technology). Seoul: T'aehaksa, 2003.

Yi Ŭn-sŏng 이은성. *Sosŏl "Tongŭi pogam"* 소설동의보감 (*Tongŭi pogam*, the novel). Seoul: Ch'angjak kwa pip'yŏngsa, 1990.

Yim Chae-hwan 임재환, Kim Chong-u 김종우, and Hwang Ŭi-wan 황의완. "Hanŭihakchŏk hwatpyŏng ch'iryo e ttara nat'ananŭn hwatpyŏng hwanja ŭi sŭt'ŭresŭ chigak chŏngdo wa imsang yangsang ŭi pyŏnhwa e taehan pigyo yŏn'gu" 한의학적홧병치료에따라나타나는홧병환자의스트레스지각정도와임상양상의변화에대한비교연구 (A comparative study on the changes of the clinical appearances and stress perception of *hwabyung* patients, according to the Oriental medical therapy). *Tongŭi sin'gyŏng chŏngsin kwahak hoeji* 11, no. 1 (2000): 47–57.

Yŏnse taehakkyo ŭihaksa yŏn'guso 연세대학교의학사연구소, ed. *Hanŭihak, sing-minji rŭl alt'a* 한의학식민지를앓다 (The modernization of Korean traditional medicine during the colonial period). Seoul: Akanet, 2008.

Yoo, Theodore Jun. *The Politics of Gender in Colonial Korea: Education, Labor, and Health, 1910–1945.* Berkeley: University of California Press, 2008.

Yu Ir-han Chŏn'gi P'yŏnjip Wiwŏnhoe 유일한전기편집위원회. *Nara sarang ŭi ch'am kiŏbin, Yu Ir-han* 나라사랑의참기업인유일한 (Yu Ir-han, the patriot and trustful businessman). Seoul: Yuhan yanghaeng, 1995.

Yun Yong-sŏp 윤용섭. *Algo namyŏn swiun hwabyŏng haesŏl* 알고나면쉬운화병해설 (A layman's guide to *hwabyŏng*). Seoul: Ŭisŏngdang, 2007.

Zhuang Zhaoxiang. *Bencao yanjiu rumen* 本草研究入門 (An introduction to *bencao* study). Hong Kong: Zhongwen daxue chubanshe, 1983.

Index

Page numbers for figures are in italics.

Harvard East Asian Monographs
(most recent titles)

Harvard East Asian Monographs

Harvard East Asian Monographs

Harvard East Asian Monographs